STEM
IS FOR
EVERYONE

Strategies for Engaging **Multilingual Learners** in the K–12 Classroom

DARLYNE de HAAN

Solution Tree | Press
a division of
Solution Tree

Copyright © 2025 by Solution Tree Press

Materials appearing here are copyrighted. With one exception, all rights are reserved. Readers may reproduce only those pages marked "Reproducible." Otherwise, no part of this book may be reproduced or transmitted in any form or by any means (electronic, photocopying, recording, or otherwise) without prior written permission of the publisher.

AI output featured in figure 4.12 generated with the assistance of Magic School.
AI output featured in figure 6.3 generated with the assistance of ChatGPT.
AI outputs of student images featured in figures 3.2, 4.4, 4.5, 5.1, and 6.15 generated with the assistance of Midjourney.

555 North Morton Street
Bloomington, IN 47404
800.733.6786 (toll free) / 812.336.7700
FAX: 812.336.7790

email: info@SolutionTree.com
SolutionTree.com

Visit **go.SolutionTree.com/EL** to download the free reproducibles in this book.

Printed in the United States of America

Library of Congress Cataloging-in-Publication Data

Names: De Haan, Darlyne, author.

Title: STEM is for everyone : strategies for engaging multilingual learners in the K-12 classroom / Darlyne de Haan.

Description: Bloomington, IN : Solution Tree Press, 2024. | Includes bibliographical references and index.

Identifiers: LCCN 2024016277 (print) | LCCN 2024016278 (ebook) | ISBN 9781960574060 (paperback) | ISBN 9781960574077 (ebook)

Subjects: LCSH: Science--Study and teaching--Social aspects. | Technology--Study and teaching--Social aspects. | Engineering--Study and teaching--Social aspects. | Mathematics--Study and teaching--Social aspects. | Multilingual education. | Culturally-relevant pedagogy.

Classification: LCC Q182.8 .D4 2024 (print) | LCC Q182.8 (ebook) | DDC 372.35--dc23/eng/20240630

LC record available at https://lccn.loc.gov/2024016277

LC ebook record available at https://lccn.loc.gov/2024016278

Solution Tree
Jeffrey C. Jones, CEO
Edmund M. Ackerman, President

Solution Tree Press
President and Publisher: Douglas M. Rife
Associate Publishers: Todd Brakke and Kendra Slayton
Editorial Director: Laurel Hecker
Art Director: Rian Anderson
Copy Chief: Jessi Finn
Production Editor: Paige Duke
Copy Editor: Jessica Starr
Proofreader: Elijah Oates
Text and Cover Designer: Kelsey Hoover
Acquisitions Editors: Carol Collins and Hilary Goff
Content Development Specialist: Amy Rubenstein
Associate Editors: Sarah Ludwig and Elijah Oates
Editorial Assistant: Anne Marie Watkins

I dedicate this book to my late mother, the always inspiring Edith C. "Edie" Oliver. Thank you for letting me make a mess on the steps on the side of the house. You will always be my greatest cheerleader, and I heard your cheers from heaven during this book-writing journey. I miss you, Mom, and I love you.

Acknowledgments

To Rafael, my ever-patient and understanding husband, who has endured all my crazy and outlandish ideas in the name of educating students so they can hopefully avoid the many hardships I encountered growing up as a minority female in a not-so-colorblind world. Thank you, Poppy, for always encouraging me to follow my love of science and mathematics and my dream of making science a wonderful and obtainable experience for all children, especially those children with fewer opportunities. I love you!

To my two wonderful children, Bremen and Lindsey, who were often the "victims" of my many science experiments, tagging along for way too many trips to Philadelphia's Ben Franklin Institute, and waking up at the wee hours of the morning to watch meteor showers (although there were times you told me you were not getting up). I love you two to infinity and beyond!

To Dr. John Kellmayer, a brilliant and encouraging professor, who encouraged me to write a book when I had no faith that I could accomplish such a huge undertaking. Thank you!

To Dr. Ayanna Cooper, the most unselfish person I have ever met. Thank you for supporting me when I called you up and said, "Hey, I'm not sure if you remember me, but we met when you were consulting at my school district. Would you mind giving me a few minutes to talk about some ideas I have?" That was the beginning of our friendship, and I am utterly grateful to you and everything you have done for me!

With heartfelt gratitude to Hilary Goff, my Solution Tree editor. Your courageous conversations, encouragement, and patience with my midnight epiphanies were so appreciated.

And as always, to my girlfriends—Rhuby, Saudia, Dr. KayJay, and Walfall—and my cousin, but more like a sister, Sharon, who supported, encouraged, and celebrated with me along this crazy ride of writing a book!

Solution Tree Press would like to thank the following reviewers:

Lindsey Bingley
Literacy and Numeracy Lead
Foothills Academy Society
Calgary, Alberta, Canada

Amber Gareri
Instructional Specialist, Innovation
 and Development
Pasadena ISD
Pasadena, Texas

Amy Kochensparger
High School Science Teacher,
 Science Department Chair
Eaton High School
Eaton, Ohio

Erin Kruckenberg
Fifth-Grade Teacher
Jefferson Elementary School
Harvard, Illinois

Nicole McRee
Science, STEM, Wellness Specialist
KCSD#96
Buffalo Grove, Illinois

Demetra Mylonas
Education Researcher
Headwater Learning Foundation
Calgary, Alberta, Canada

Melisha Plummer
Assistant Principal
Atlanta Public Schools
Atlanta, Georgia

Visit **go.SolutionTree.com/EL** to
download the free reproducibles in this book

Table of Contents

ABOUT THE AUTHOR ix

INTRODUCTION 1
    Current Challenges 2
    My Story 4
    About This Book 6

1 Empowering Multilingual Learners Through STEM Education ... 9
    Barriers to Closing the Achievement Gap 12
    Implications for Schools and Teachers 16
    Key Takeaways 22

2 Understanding Multilingual Learners' Unique Needs ... 25
    Multilingual Learners' Assets 26
    Factors Unique to Their Experience 27
    STEM as a Solution 38
    Key Takeaways 38

3 Making Instruction Applicable Through Culturally Responsive Teaching ... 41
    Culture Informs Relationships 44
    Culture Informs Behavior 45
    Implicit Bias as a Barrier 47
    Meet Maria 48

Culturally Responsive Teaching in Practice 52
The STEM Challenge 58
Key Takeaways 74

4 Using Collaborative Learning Groups to Support Language Acquisition and Sustain Rigor 77

Cooperative Learning Models 78
Two Pairs in a Quad 84
Meet Jean Pierre, Linh, and Amihan 88
Collaborative Learning Groups in Practice 93
The STEM Challenge 102
Key Takeaways 109

5 Leveraging Student Assets and Building Content Knowledge Through Scaffolding 111

Terms Defined 113
Funds of Knowledge and Instruction 114
Meet Fatou 121
Scaffolds in Practice 122
The STEM Challenge—Introduction 134
The STEM Challenge 144
Key Takeaways 153

6 Using Claim, Evidence, and Reasoning to Build Language Fluency 155

CER's Four-Part Framework 156
How CER Helps Multilingual Learners Build Language Proficiency . . 158
CER in Your Classroom 159
CER in Action 163
Meet Wang Xiu Ying 175
CER STEM Challenge 176
Key Takeaways 178

EPILOGUE 181

REFERENCES AND RESOURCES 183

INDEX 197

About the Author

Darlyne de Haan, Ed.D, is the director of Curriculum, Instruction, Innovation and STEAM for a school district in southern New Jersey. She is a former forensic scientist for the state of New Jersey; a former environmental chemist; and instructional specialist for the New Jersey Department of Education, where she provided professional development to over thirty school districts in the state labeled as either in need of improvement or failing. Dr. de Haan is an often-sought-out international and national speaker on issues related to English learners and minority students of low-income communities in STEM learning areas. She is the founder and executive director of Mad About Science Learning Center, Inc., a 501(c)(3) nonprofit organization under which she created two platforms, Neighborhood-Science and Brain Based Science, that provide high-quality, educational, and kid-friendly science and educational videos to explain the brain's impact on social-emotional development, instruction, and other learning issues affecting multilingual learners and other vulnerable populations.

Dr. de Haan is a proud recipient of the coveted Fulbright International Education Administrators Program award for Fulbright Leaders for Global Schools, a program sponsored by the U.S. Department of State's Bureau of Educational and Cultural Affairs. She is part of a cohort that traveled to Singapore to partake in the International Education Administrators seminars. The program helps U.S. international education professionals and senior higher-education officials create empowering connections with other countries' societal, cultural, and higher-education systems with a goal to empower students to become global citizens.

Dr. de Haan received a bachelor's degree in chemistry, biochemistry, and mathematics from Douglass Residential College, part of Rutgers University; master's degrees from Central Michigan University and Stockon University; and a doctorate in organizational leadership, specializing in multilingual learners, from Stockton University.

To learn more about de Haan's work, follow @de_darlyne68474 on X, formerly known as Twitter.

To book Darlyne de Haan for professional development, contact pd@SolutionTree.com.

Introduction

> *Our vision is to ensure every multilingual student engages in learning that allows them to strive academically and choose their path for success.*
>
> **—ENGLISH LEARNERS SUCCESS FORUM**

In a 2014 *Education Week* article, managing editor Lesli Maxwell (2014b) reported that, during the 2014–2015 school year, the United States' K–12 school population would enter a new era in which White students would no longer constitute the majority of public-school children. The National Center for Education Statistics' projections also predicted that the collective number of Hispanic, African American, Asian, and Native American students would outnumber their White peers (Maxwell, 2014b).

In the 21st century, we are now seeing firsthand that the projections are proving to be correct and are affecting many school districts around the United States (Ferlazzo, 2022). Yet, teacher education programs, mindsets, and general attitudes about preparing future teachers for this dynamic cultural shift remain stagnant, even with so much advance notice (Barone-Crowell, 2020).

All students need to become critical thinkers, which great science, technology, engineering, and mathematics (STEM) instruction can foster. However, many people confuse STEM as being just a course or falsely believe that a project or task is not truly STEM if it is not explicitly using technology, mathematics, and engineering. What this outlook fails to realize is that projects don't have to include all STEM components to be "STEM projects." All that's required is that some of those components were used to come up with the project's solution.

STEM is not a course. It is a way of thinking critically. It is a way to solve problems using knowledge from one or all four areas of the STEM acronym combined with a student's funds of knowledge (see chapter 5, page 111). This is why it is important that multilingual learners be allowed to enroll in STEM courses—it provides them with the opportunity to simultaneously learn content, language, and problem-solving and critical-thinking skills. This book's purpose is to provide guidance and tools for teaching STEM to multilingual learners.

In this introduction, I'll discuss current challenges that public education faces in meeting multilingual students' needs, share the story of how I overcame personal challenges in public education that led me to advocate for multilingual learners, and describe this book's intention as a comprehensive resource for K–12 teachers educating multilingual learners in STEM content. This introduction ends with guidance about how to use this book, including an overview of each chapter.

Current Challenges

This student population has received different labels over time. They are sometimes referred to as *English learners*. This book uses the term *multilingual learner* to reflect the reality that many of these students are not just learning English as a second language but may be learning their second, third, or fourth language. Educators face distinct challenges in meeting multilingual learners' unique needs. This section will look at these challenges in more detail.

Many multilingual learners struggle because they have limited access to quality instruction tailored to their specific needs. Many students enter the classroom without proficiency in the language of instruction, and many content-area teachers enter the classroom feeling unprepared to teach language learners (McGraner & Saenz, 2009). If mainstream teachers are expected to create inclusive classrooms where multilingual learners receive the instructional support they need and deserve, then some targeted intervention is needed. These limiting beliefs must change; mainstream teachers must see multilingual students as capable learners and themselves as responsible for these students' learning.

The growing presence of multilingual learners in U.S. schools is provoking change in how multilingual learners are educated and why education can no longer be business as usual for this student population. We can no longer use language proficiency as a prerequisite for admittance into rigorous-content or STEM-related classes; growing research demonstrates that multilingual learners can be successful in STEM classes with limited English proficiency, and they can learn content and language simultaneously (National Research Council, 2000).

But one of the factors keeping multilingual learners out of STEM classrooms and careers—aside from strict language proficiency requirements for selection into STEM classes—is the fact that most multilingual learners live in a state of poverty. STEM fields have the highest success of changing lives and livelihoods; with over 60 percent of multilingual learners living in poverty, how can we, as educators, not see the importance of having every multilingual learner enrolled in a STEM program, class, or summer program (Breiseth, 2015)? For many multilingual learners, schools are their only chance to experience STEM because their parents—due to structural barriers, such as lack of transportation, program fees, and English language prerequisites—cannot afford these potentially life-changing opportunities for their children.

Linguistic diversity is increasingly characteristic of today's classrooms. As the Hispanic population continues to grow—and to grow up—so, too, will the number of students experiencing learning difficulties in the classroom, unless educators design instruction to match today's student demographics (Lesaux, 2013). *Education Week* reporter Corey Mitchell (2017) argues that teacher and principal preparation programs must adapt their curricula and training to ensure that all present and future educators are ready to work with an ever-diversifying student population. School districts must also evaluate current teachers' needs in supporting multilingual students. To meet future needs, teacher preparation programs must prepare educators for the projected population of multilingual learners. Mitchell (2017) proposes that national coordination of these efforts would be most effective to prepare both preservice teachers and those currently in classroom placements because most teacher education programs do not train the general education teacher how to teach diverse populations or students whose first language is not English. For this type of instructional preparation, teacher candidates must select the academic track to teach students from urban communities, called *urban education*, or the track for students who do not speak English.

One troubling implication of this lack of training is negative attitudes toward multilingual learners and their needs. Some teachers have varying attitudes about multilingual learners and their inclusion in mainstream classrooms (Barone-Crowell, 2020). Education researcher Jenelle R. Reeves (2006), in a survey of 279 content-area teachers, finds that 90 percent had received no training to work with language-minority or multilingual students. Findings also show negative attitudes in the teachers surveyed. Forty percent of respondents do not believe that all students—multilingual learners and general education students—benefited from multilingual learners' inclusion in the mainstream classroom. In addition, the findings show negative attitudes toward specific types of coursework modification for multilingual learners. Fifty-four percent of respondents disagree with modifying coursework for multilingual learners. Furthermore, approximately 70 percent of the teachers report that they do "not have enough time to deal with the needs of ESL students" (Reeves, 2006, p. 136).

Another challenge for teachers of multilingual learners is the creation of the Next Generation Science Standards (NGSS), also referenced as the Next Gen Standards, which were released in April of 2013 and adopted in many U.S. states. The NGSS identifies science and engineering practices and content that all K–12 students, including multilingual learners, should master to be prepared for college, careers, or citizenship (Miller, n.d.). With the NGSS, or versions of it, being used in almost every U.S. state, teachers face a shift from more traditional teaching methods to a more inquiry-based and hands-on learning approach.

The NGSS are not traditional standards, such as those typically found in content areas like English language arts and mathematics. Generally, those standards are direct and require a skill to be demonstrated or explicit knowledge to be learned. These work well in an assessment-centered environment. In the era of high-stakes testing at the start of the century, teachers could improve test scores by teaching directly to these standards. The NGSS authors set out to create something different. The standards are built to support not just the skills and content that students need for science, but also the overarching concepts that cannot be demonstrated through a multiple-choice assessment. The standard's focus lies heavily on abilities, such as developing and using models, arguing from evidence, and analyzing data:

> As an example, consider a common second grade lesson developed around the standards on "Why is our corn changing?" Children learn that harvest corn, something initially seen as decoration, suddenly begins to sprout what appears to be leaves and roots. Disagreements about how the corn is growing then

spark a series of questions and ideas for investigations related to what is causing this growth. So rather than focus on mere reasons for the change, as might have been done traditionally, the students are taught instead how to assess and determine the information that will be most helpful to them when determining the cause—a perspective that will serve them well when facing other situations requiring problem-solving. (Vigeant, 2021)

Teachers are encouraged to adopt a broader perspective in their instructional methods when teaching with NGSS's three-dimensional learning approach (disciplinary core ideas, science and engineering practices, and crosscutting concepts). These standards challenge teachers to make this paradigm shift of describing phenomena, collecting data, and making evaluations. This new approach to teaching science should foster a learning approach that is applicable in real-world situations. Students should perceive their learning experiences as holistic and interconnected. Even so, integrating disciplines and interleaving the three dimensions of NGSS is complex for teachers to combine with existing curriculum. These challenges make it troubling or problematic for some educators to implement NGSS. Crosscutting concepts might be the biggest challenge for teachers because they require a mindset shift toward using technology, computational thinking, and engineering design concepts in teaching (Vigeant, 2021).

Considering all of these challenges, you can understand why professional learning should play a larger role in preparing educators to teach multilingual learners. My research over the years has shown that teacher textbooks still only contain a vague paragraph with recommendations for accommodating multilingual learners. Teachers aren't receiving the necessary training to make these more-complex lessons accessible for multilingual learners.

This book provides strategies, processes, and scaffolding ideas for teaching all students STEM, especially those whose first language is not English. Not only is it the correct thing to do, but multilingual learners are also an untapped resource for developing a multilingual STEM workforce that is an asset in an increasingly competitive global economy. However, this is not the case in many school districts, where multilingual learners are too often perceived as academic underachievers according to the U.S. Department of Education Office of English Language Acquisition, *High-Quality STEM Education for English Learners* (Shi, 2017). With the growth of multilingual learners outpacing the overall growth of preK–12 student populations, multilingual learners can be an important target group to increase the STEM workforce.

For these reasons, far too many future teachers still lack the skill set and best practices for effectively teaching minority students and multilingual learners STEM and other rigorous courses. Public education is not prepared for the inevitable cultural change already in progress.

My Story

I have both a personal and professional interest in this topic. As a Black, female youngster growing up in a single-parent household with dreams of being a scientist, I experienced firsthand how students are affected by teachers' outdated paradigms. My teachers' mindsets had grave implications for my educational journey, and as a teacher, I continue to witness diminished educational opportunities for students whose first language is not English.

I have always had a curious mind and a love for science and mathematics. My mother supported my curiosity even when it meant I was making a mess around the house. As a child, I loved combining household ingredients to see what I could create. Because we couldn't afford Barbie accessories, I made them myself using flour and water. I mixed the ingredients together until I got the right consistency; added food dye; formed the paste into plates, cups, or a makeshift car; and let my creations dry in the sun. I designed and created Barbie's clothes using fabric scraps from my mother's sewing room and used empty shoe boxes to create a multilevel house. My imagination was endless, and I knew early on that I wanted to be a scientist when I grew up. Not a Black scientist or a female scientist, just a scientist.

In the early 1970s, after moving at the age of twelve, mine was the only African American family in my small northern New Jersey community, and I was the only African American child in the school district. I experienced many microaggressions in my academic and social life in my new school district.

Though I did well in school and loved science and mathematics, many of my White middle-class teachers held low academic expectations for me and seemed unprepared to work with minority students or provide culturally responsive teaching. My teachers seemed to experience cognitive dissonance as they witnessed me excel at mathematics and science. My high school and college years followed the same pattern.

After graduating from college, I achieved my childhood dream and became a forensic scientist. I promised myself I would never forget how teachers' low academic expectations affected me or what it was like to work harder just to be on par with my White peers. It was these memories that eventually motivated me to become a teacher and make a difference in the lives of minority students who loved mathematics and science.

When I became a teacher and earned a master's degree in education, I saw more clearly how a lack of training and professional development left many teachers unprepared to work effectively with culturally diverse students. My research and writing interests expanded to include speakers of languages other than English in the general education classroom.

Initiatives I've launched regarding this work follow.

- Webinars and workshops on topics including teaching with the brain in mind and understanding language acquisition
- A YouTube channel on educational topics like instructional strategies for teaching multilingual learners STEM and how the English learner's brain learns
- Two websites providing resources for educators
- Professional development for school districts on how to teach inquiry-based, hands-on science
- After-school and summer science programs for multiple school districts

- A learning center to help students improve their mathematics abilities
- Grants from major community organizations and industries to work with minority students to learn about STEM and STEM careers
- A nonprofit organization called Mad About Science Learning Center, Inc., with a mission to change the face of STEM by training teachers how to teach inquiry-based science and mathematics.

You can access these resources by scanning the following QR codes. My experience as a child, student, teacher, and advocate in these diverse contexts informs my approach to this book.

brainbasedscience.com

neighborhood-science.com

@dr.ddarlynedehaan4401

About This Book

This book aims to provide a comprehensive resource of STEM content fields for general education teachers. *All* students, regardless of their home language or racial affiliation, should receive a fair and equitable STEM education. For this to happen, general education STEM teachers must understand the language acquisition process (see table 2.1, page 30). They also need to understand language acquisition theory. According to American linguist and educator Stephen Krashen (n.d.), students acquire language through comprehensible input—language that is one step beyond their current language proficiency level. And, most importantly, these teachers need to know why we are all teachers of language and what that looks like in a STEM content classroom.

It's true that every teacher is an English teacher. This doesn't mean that every teacher needs to be a language expert—rather they are experts of their subject areas' language, and they provide extensive support to aid student comprehension and participation (Mohamed, 2024). This book is written for you, the general education teacher. Maybe you've never completed an English language learning course. Maybe you're alternatively certified and never encountered coursework on teaching multilingual learners. The student scenarios, scaffolds, recommendations, and real-life examples in these pages are more than a paragraph on the border of your teacher guide. You will walk away from this book with a new sense of self and self-efficacy when teaching multilingual learners STEM content. You will learn how to teach STEM to students whose native language is not English by building on students' existing language resources and knowledge. Educators must expand students' basic linguistic repertoire and simultaneously build their academic language by using their previous knowledge and developing their new knowledge of STEM in another language.

Antiquated teacher education programs that separate coursework for diverse student populations to special certificate tracks cannot continue. And schools cannot continue to believe that a few hours of professional learning will provide general education teachers with the skills and knowledge needed to successfully teach multilingual learners language-rich STEM courses. This book intends to show it is possible and beneficial to meet multilingual learners' unique needs in the general education classroom. And finally, this book provides teachers with best practices on teaching multilingual learners in a mixed language-proficiency classroom and aims to increase teachers' self-efficacy in teaching students whose first language is not English.

This book is intended to be your instructional guide as you implement successful teaching strategies for multilingual learners in your STEM classroom. Each chapter includes QR codes that link to supplemental content online. You'll also encounter strategies, activities, templates, tools, and lesson plans to aid your journey to supporting multilingual learners. By the end of the book, you will have a better understanding of how to address a variety of student situations, how to design lessons according to different language acquisition levels, how to be specific and targeted in creating student groups based on academic and language ability, and best of all, you'll learn how to make STEM challenges appropriate for all students.

Chapter 1 discusses STEM's potential to empower multilingual learners and the steps public education must take to stop this demographic's disproportionate exit from STEM education.

Chapter 2 explores multilingual learners' unique needs, including language acquisition and proficiency, altered instructional design, additional time, and scaffolding.

Chapter 3 describes how teachers make instruction applicable to multilingual learners through culturally responsive teaching practices, such as developing cultural knowledge, valuing students' diverse cultures, and identifying implicit bias. The chapter includes a STEM challenge incorporating culturally relevant practices.

Chapter 4 explains why cooperative learning is ideal for multilingual learners and introduces a grouping strategy called Two Pairs in a Quad. This method allows multilingual learners to work in diverse group configurations that maximize exposure to linguistic and academic content. The chapter includes a STEM challenge incorporating Two Pairs in a Quad.

Chapter 5 illustrates what teachers can do to leverage multilingual learners' assets, tap their unique funds of knowledge, and build content knowledge through scaffolding. The chapter includes a STEM challenge incorporating scaffolding.

Chapter 6 introduces the three-part claim, evidence, and reasoning (CER) framework as a tool to support multilingual learners in building language proficiency. The chapter uses two examples to model how teachers can use CER in their classroom.

It is my hope that this book provides everything you need to make the biggest difference in the lives of your multilingual learners. Though they may not speak English well *yet*, your belief in their potential to excel in the English language and in STEM is paramount to their success both inside and outside of the classroom.

CHAPTER 1

Empowering Multilingual Learners Through STEM Education

> *It is my job to help all my students belong . . . this means part of my job is to educate the wider community about the needs of a child who may be perceived as "different" to help them become embraced as part of the community.*
>
> **—GAYLE HERNANDEZ**

In the introduction (page 1), you learned that multilingual learners continue to fall behind their English-speaking peers. Although this population's academic underperformance is not a new phenomenon, this issue came to the forefront with the passage of the Every Student Succeeds Act in 2015, when federal law, for the first time, held schools accountable for multilingual learners' academic achievement in all content areas and progress toward developing English proficiency (Cooper, 2021). Unfortunately, the Every Student Succeeds Act hasn't changed student performance. Marginalized culturally and linguistically diverse students are still struggling academically and performing far below their White, English-speaking peers, especially in mathematics and science.

Figure 1.1 (page 10) illustrates how the gap in average National Assessment of Educational Progress (NAEP) science scores between multilingual learners and their English-speaking peers remained unchanged from 2009 to 2015.

> **FACT:**
>
> Average NAEP science scores for English learners in fourth and eighth grade increased by seven points between 2009 and 2015. Average scores for students who are not English learners increased by four points between 2009 and 2015.
>
> There was a thirty-six-point gap between the average scores of English learners and not English learners in fourth grade in 2015. There was a forty-six-point gap between the average scores of English learners and students who are not English learners in eighth grade in 2015.

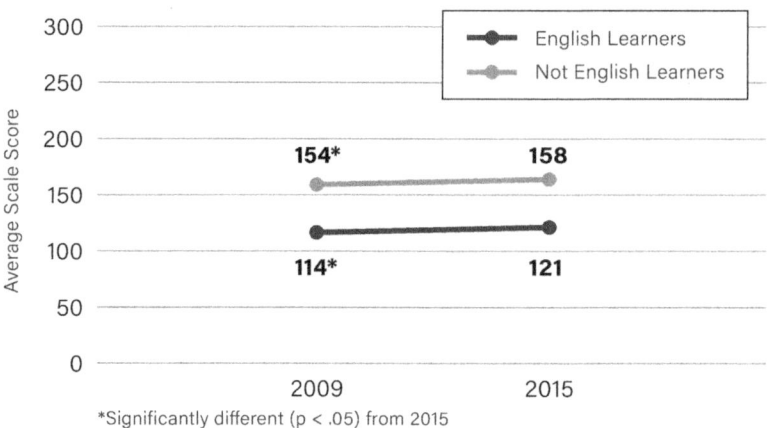

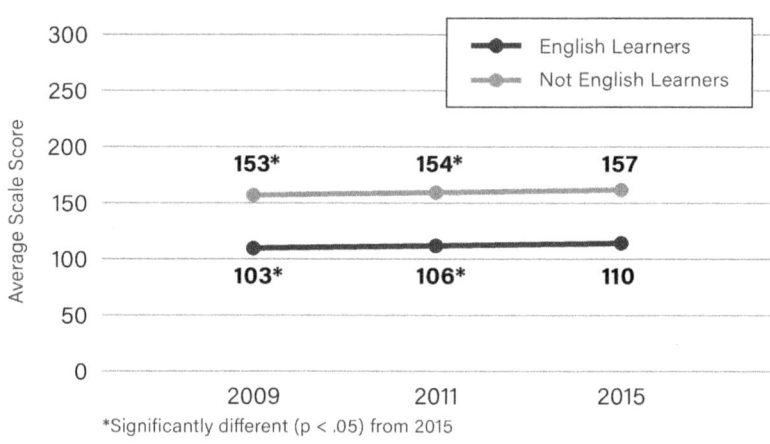

Source: Nation's Report Card, 2019a and 2019b.

FIGURE 1.1: NAEP science score comparison for multilingual learners, grades 4 and 8.

Figure 1.2 illustrates the disparity between multilingual learners and their English-speaking peers enrolled in high school mathematics classes.

Physicist and materials science researcher Julia Phillips writes, "Our economy depends on mathematics and science literacy . . . What we see is that the performance of children in the U.S. has not kept pace with the performance of students from other countries in science and mathematics for a decade or more" (as cited in Gillespie, 2021). U.S. students' scores are last place in mathematics and in the middle range in science when compared with the closest economic competitors (Gillespie, 2021).

This is compounded by the rapid growth of multilingual learners in the United States. According to The Civil Rights Project, nearly all teachers across the United States can expect to have multilingual learners in their classroom (Santibañez & Gándara, 2018). William H. Frey (2018a), author of *Diversity Explosion: How New Racial Demographics Are Remaking America*, writes that the rapidly growing Hispanic, Black, Asian, and multiracial American

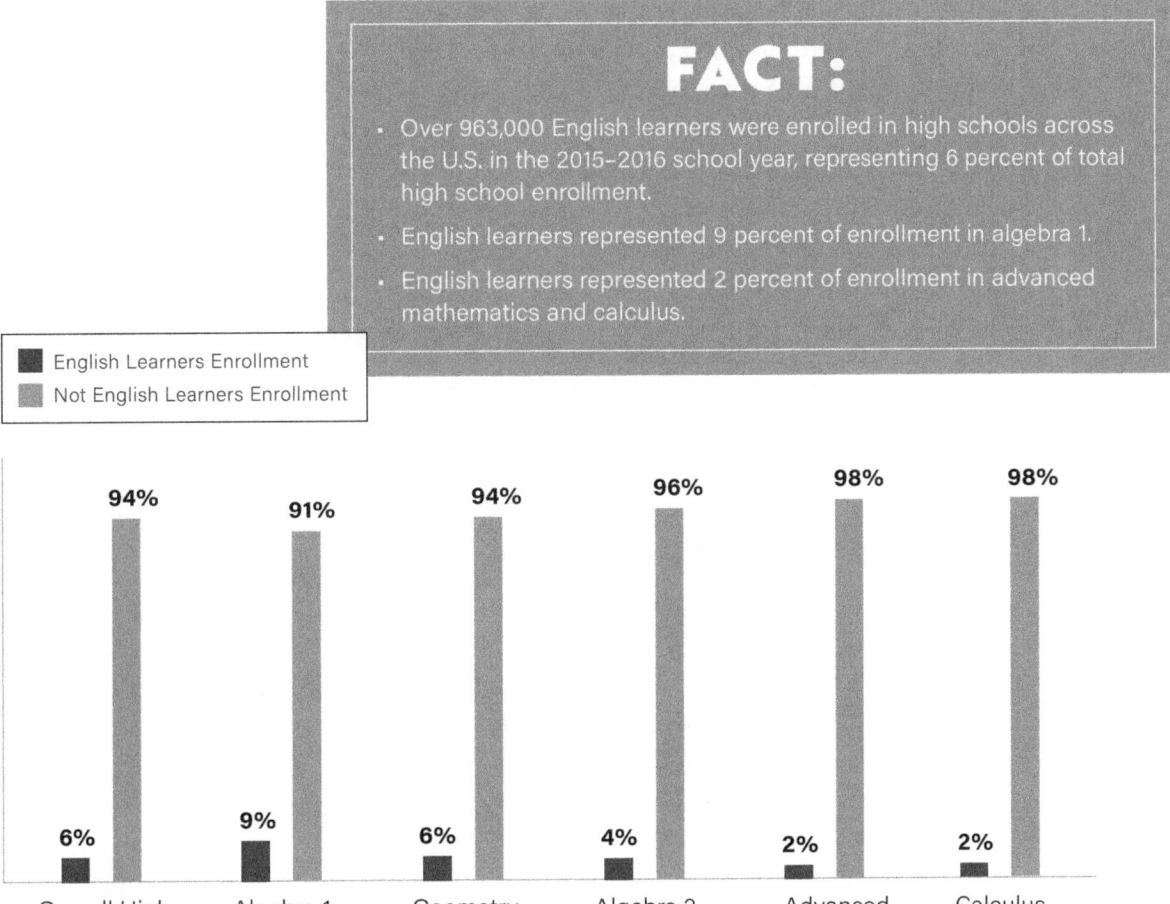

Source: *Office for Civil Rights, 2018, p. 9.*

FIGURE 1.2: Distribution of students enrolled in high school mathematics classes.

populations are transforming and reinvigorating the United States' demographic landscape. Youth and working-age populations, along with voters and consumers, will primarily be fueled by minorities in the foreseeable future (Frey, 2018b). Consequently, the predominantly White senior population will rely more on their input to the economy and government programs like Medicare and Social Security. This minority growth underscores the importance of sustained investment in diverse youths and young adults as the population ages.

Maxwell (2014b) reports that, by 2050, 34 percent of children younger than seventeen in the United States will either be immigrants themselves or the children of at least one parent who is an immigrant, according to projections from the Pew Research Center. That portends an ever-growing share of children coming to schools with few or no English language skills. Educators and public policymakers must heed these demographic projections and plan for the increasing need for English language instructional programs. Public education must train teachers to not only deliver solid core-content instruction, but also English language development and literacy instruction.

In this chapter, I discuss these barriers to closing the achievement gap between multilingual learners and their English-speaking peers—the leaky science, technology, engineering, and mathematics (STEM) pipeline; inadequate teacher preparation; admission requirements; and socioeconomic barriers. If public education is going to ensure multilingual learners succeed, it must commit to sealing the leaky STEM pipeline through increasing teacher self-efficacy, ensuring proper teacher preparation, and granting multilingual learners access to STEM education beginning in elementary grades.

Barriers to Closing the Achievement Gap

Barriers to closing the achievement gap in education typically refer to the challenges that prevent equal educational opportunities and outcomes among students from different backgrounds. These barriers can significantly impact students' academic performances, future opportunities, and socioeconomic mobility. The following sections discuss some of the key barriers impacting multilingual learners' access to STEM opportunities.

The Leaky STEM Pipeline

Students need to follow a certain educational path beginning in early education (elementary age) and continuing to secondary education (high school age) if they want a better chance of having a career in a STEM field. This path is often called the STEM pipeline. Students need sufficient experience with topics in the field and need to retain that learning as they complete their education in order for a sufficient output of graduates to make up the workforce (Sass, 2015). The leaky STEM pipeline describes students' disproportionate exit from participation in these content areas throughout K–12 and college, resulting in their underrepresentation in STEM careers. Middle and high school multilingual learners, students of color, and girls—particularly those from low-income families and schools—are disproportionately excluded or dropped from the STEM pipeline at formative moments in their academic trajectories, specifically the middle school years, ages 12–14 (Lyon, Jafri, & St. Louis, 2012). For most culturally and linguistically diverse students, school is often their only opportunity for exposure to STEM and rigorous content. School, therefore, becomes these students' sole opportunity to break the cycle of poverty and open possibilities of successful employment and careers in STEM.

Inadequate Teacher Preparation

As you read in the introduction (page 1), teacher preparation is a major factor in multilingual learners' success. Given the current demographic shifts in the U.S. population, it is likely that all teachers, at some point in their careers, will encounter students who do not yet have sufficient English proficiency to fully understand the academic content in the mainstream classroom (de Haan, 2019). Many teachers are not prepared to provide high-quality instruction to these student populations (Ballantyne, Sanderman, & Levy, 2008). Additionally, education researchers Thi Diem Hang Khong and Eisuke Saito (2014) note that "60% of deans of colleges of education admitted to the lack of adequate focus on [multilingual learners' needs] in coursework of their teacher education programme" (p. 214).

Lack of teacher preparation is influenced by current legislation that affects multilingual learners' education, deficits in mainstream teacher preparation programs, and deficits in multilingual learners' professional development. These issues, when combined, negatively affect mainstream teachers' abilities to educate multilingual learners in their classrooms. These issues can cause mainstream teachers to be unprepared as well as hold negative attitudes toward educating multilingual learners. According to educator Rose G. Skepple (2015), teacher education programs must assist preservice teacher candidates in critically examining their views about diversity, their expectations about teaching in diverse settings, and their responsiveness to student differences.

Even good teachers may find it difficult to meet multilingual learners' needs without special preparation (Santibañez & Gándara, 2018). Multilingual learners have unique needs that mainstream teachers are not equipped to meet in their degree path. The lack of teacher preparation to effectively teach this growing population will continue to profoundly impact multilingual learners' ability to receive an equitable STEM education.

Admission Requirements

Many school districts have unrealistic requirements in place for admittance into rigorous courses like STEM. This is problematic given that these courses have been called "the gateway out of poverty" for many multilingual learners because STEM jobs have the potential to transform students' lives (Change the Equation, 2017). According to David Francis and Amy Stephens (2018):

> English learners (ELs) bring a wealth of resources to science, technology, engineering, and mathematics (STEM) learning, including knowledge and interest in STEM-related content that is born out of their experiences in their homes and communities, home languages, variation in discourse practices, and, in some cases, experiences with schooling in other countries. ELs are those students ages 3 through 21, enrolled in an elementary or secondary school, not born in the United States or whose native language is a language other than English, and whose proficiency in speaking, reading, writing, or understanding the English language may be sufficient to deny the individual the ability to successfully achieve in classrooms where the language of instruction is English. (p. 1)

Jobs in STEM fields are on the rise. These jobs offer tremendous economic opportunities and hold promise for transforming the lives of individual students, their families, and society at large (Muñiz, 2019). According to the National Academy of the Sciences report compiled by David Francis and Amy Stephens (2018), too often schools operate under the incorrect assumption that English proficiency is a prerequisite to meaningful engagement with STEM learning. This assumption, which research disproves (Sullivan, 2019), fails to leverage multilingual learners' meaningful engagement with content and disciplinary practices as a route to language proficiency. Despite rapidly increasing demand for STEM workers, which is outpacing job growth in the United States, multilingual learners are largely being left out of the STEM revolution (Funk & Parker, 2018).

Lack of interest or ability are not obstacles to English learners participating in STEM subjects and careers; rather, it stems from a lack of access to rigorous, age-appropriate instruction in the field (Sullivan, 2019). The beauty behind STEM is that it's an effective tool for multilingual students to learn both content and language. STEM subjects include alternative routes to acquiring knowledge like experimentation, demonstration of phenomena, and demonstration of practices through which students can gain a sense of STEM content without resorting mainly to language to access meaning (Francis & Stephens, 2018).

Because STEM learning is core to many new technologies and jobs, it is important that schools connect students with this material earlier, and in a more meaningful way, and stop using English level proficiency as an entrance requirement. Research suggests that a shift is needed; educators and educational systems must recognize the assets that multilingual learners bring to the classroom and understand that some deficits in student performance arise from lack of access rather than limited ability, language proficiency, or cultural differences (Harper, 2019).

Socioeconomic Barriers

Socioeconomic status is another contributing factor to the multilingual learners' experience in the U.S. public education system and it exacerbates the leaky STEM pipeline. Teach for America editorial director Jessica Fregni (2021) reports that 60 percent of multilingual learners in the United States live in low-income communities and often attend schools with high teacher turnover. Students from low-income families living in low-income communities suffer from pervasive structural barriers that exclude them from participating in STEM out-of-school opportunities, such as summer camps and programs in nearby communities. The following are some barriers that have been associated with the lack of STEM opportunities (Lyon, Jafri, & St. Louis, 2012).

- Inability to pay program registration fees
- Lack of prerequisite knowledge
- Highly competitive application processes
- Inability to demonstrate preexisting interest in science
- Poor literacy skills
- Lack of transportation
- A dearth of accessible opportunities

Although research suggests education spending can improve student outcomes—especially among low-income students—students who attend the highest-poverty schools are least likely to have access to STEM resources, experiences, and classes that wealthier parents can seek out for their children (Change the Equation, 2017). And when it comes to state funding, districts with concentrated poverty are still getting far less than they need, despite decades of efforts to improve funding disparities (Katz, 2020). As a result, students in such schools suffer multiple disadvantages over the course of their schooling, and they face dim prospects for rewarding STEM careers (Change the Equation, 2017).

However, state and education leaders can adopt proven policies and strategies for boosting opportunities in schools with the highest concentrations of poverty despite decades of economic segregation, which are difficult to undo through education policy alone (Change the Equation, 2017). These data are echoed by the National Science Foundation's findings that science and mathematics performance are not equally distributed across the United States (Gillespie, 2021). Huge differences exist in performance based on race and ethnicity where it shows that Asian and White students do much better on these standardized tests than other students of color. Additionally, there is a huge difference based on students' socioeconomic background—students from higher socioeconomic backgrounds do much better than students from low socioeconomic backgrounds (Gillespie, 2021).

The Elementary and Secondary STEM education, Science and Engineering Indicators 2022 report finds the following regarding K–12 achievement in mathematics (Rotermund & Burke, 2021).

- The United States ranked twenty-fifth out of thirty-seven Organisation for Economic Co-operation and Development (OECD) nations in mathematical literacy among fifteen-year-old students.
- National mathematics test scores for U.S. minority (non-Asian) eighth graders were lower than those of their White and Asian peers.
- Middle school mathematics teachers with in-field degrees were less prevalent at high-minority-enrollment schools.
  - 75 percent of mathematics teachers had in-field degrees in schools with less than 25 percent minority enrollment.
  - 61 percent of mathematics teachers had in-field degrees in schools with greater than 75 percent minority enrollment.

A report by education researchers Dan Goldhaber, Lesley Lavery, and Roddy Theobald (2015), states that teaching in poorer schools is a particularly tough job and that, as teachers gain more experience, they use their ability to teach in an urban, low-income school as a negotiating tool to obtain employment in more advanced schools outside of urban areas. This creates high teacher turnover and an abundance of vacancies, especially in the mathematics and science courses where turnover rates are 70 percent higher for teachers in schools serving the largest concentrations of students of color (Schiff, 2023). Teachers in these schools have fewer years of experience and, often, significantly less training. According to research by Desiree Carver-Thomas and Linda Darling-Hammond (2014):

> Teacher turnover rates are 90% higher in the top quartile of schools serving students of color than in the bottom quartile for mathematics and science teachers, 80% higher for special education teachers, and 150% higher for alternatively certified teachers (p. v).

Multilingual learners face unique challenges in education, which can be exacerbated by poverty, especially in the context of STEM education. How schools address these growing projections will have huge implications for multilingual learners, who trail their English-speaking peers on most academic success measures (Maxwell, 2014b). It is urgent that United States public education provides STEM opportunities for all students. This is not only a concern for increasing the numbers of those with STEM careers, but also for the United States at large.

Implications for Schools and Teachers

If school districts continue enacting criteria and requirements based on language proficiency and prerequisite skills and knowledge, this student population will continually be omitted from participation in STEM content courses, perpetuating the leaky STEM pipeline.

Public education in the United States must seal the leaky STEM pipeline for multilingual learners and other underrepresented populations. What areas can have the greatest impact on students' exposure to STEM courses? I advocate for the following focus areas.

- General education teachers need to receive proper training for teaching diverse populations successfully.
- School districts need to reevaluate their policies and requirements for acceptance into STEM or rigorous courses.
- School districts need to focus on what students in underrepresented populations can do and the contributions these populations can bring to class discussions and perspectives, as opposed to what they cannot do.

While it's important that we acknowledge these factors, they're not under the general education teacher's control. Given that this book is for the general education teacher, we focus on three areas of improvement where school districts can move the needle: (1) teacher self-efficacy, (2) teacher preparation, and (3) early grades impact on student perception of STEM and language development. The following sections look at each of these in more detail.

Teacher Self-Efficacy

Researchers Lisa A. Ruble, Ellen L. Usher, and John H. McGrew (2011) define *teacher self-efficacy* as the "beliefs teachers hold regarding their capability to bring about desired instructional outcomes" (pp. 67). When teachers struggle to succeed in teaching multilingual learners, they begin to doubt their ability to be effective teachers for all students, not just multilingual learners—their self-efficacy takes a hit. This creates a dilemma because without specific coursework or professional development relating to multilingual learners' unique needs, general education teachers will not be able to successfully teach these students. Yet, in most classrooms across the United States, general education teachers play a critical role in multilingual learners' education (de Haan, 2019).

Teacher attrition rates are highest during their first year of service and even higher in schools with a high percentage of low-income and minority students. Lower teacher self-efficacy beliefs have also been linked to inadequate teacher preparation (de Haan, 2019). If schools want to ensure teacher retention, they must make teacher self-efficacy a top priority (Moseley, Bilica, Wandless, & Gdovin, 2014). School systems need to understand that when teachers are not provided with the knowledge, resources, and professional development to teach diverse student populations, teacher attrition rates will most likely continue to be significantly higher in the urban communities where they are needed most.

Districts are responsible for providing the ill-equipped general education teacher with ongoing professional development—not the all too familiar one-and-done professional development, where someone comes in two or three times a year and bombards staff with information without further implementation support. This manner of professional development becomes nothing more than a check-off list that contains no substance or, as my mother used to say, "no meat on the bone." All teachers, but especially general education teachers who are responsible for teaching multilingual learners, need consistent, embedded-coaching professional development. Otherwise, teachers won't develop the underlying beliefs and viewpoints needed to create better learning experiences and environments for these student populations (de Haan, 2019).

Important topics for professional development and training for teachers of multilingual students include language acquisition process, language acquisition levels and what students and teachers can and should do at each level, culturally responsive teaching, and culturally responsive pedagogy to understand the importance of critical introspection of teachers' beliefs about multilingual learners and linguistic diversity.

It's crucial that teachers are trained to identify and understand the stages of language acquisition and what students can do and understand at each stage. Without this understanding, teachers are not equipped to properly scaffold instruction for multilingual learners to be successful (IRIS Center, 2022a). This can lead to the teacher's belief that they are failing in their responsibility to teach all students or that the students do not possess the knowledge, desire, or aptitude to be successful in STEM or higher-level courses. This leaves multilingual learners to remain in remedial classes, reducing their ability to exit the school's English language instruction—often referred to as English as a second language program—or resulting in their being re-enrolled in the program (Ma, 2022). This critically affects students' opportunities to take STEM courses, and worse, their belief that they are incapable of succeeding in such careers.

In my experience as both a teacher and district administrator, teachers spend a lot of time (three to four times per school year) attending mandated, district-level professional development in addition to after-school or before-school professional development mandated by their school-level administrator. But too often, many of the professional development offerings are one-size-fits-all presentations that fail to meet their needs, leaving teachers frustrated and stressed. Teachers crave more relevant professional development—training that is personalized, efficient, and actionable. While the district's approach to professional development matters, it's outside teachers' control. What can teachers do to get the professional development they need?

Kareem Farah and Robert Barnett (2021), founders of the Modern Classrooms Project, suggest that teachers should ask their school leaders to provide professional development that takes their diverse needs into account and empowers them to support their students. My district provided teachers with professional development choices after surveying teachers about the kinds of offerings they would like for the upcoming school year. This initiative included an understanding that some district sessions may be mandatory to fulfill obligations like safety, new programs, and so on; teachers scheduled their selections around the mandatory sessions. This was well received because the teachers were able to find relevant professional development. The sessions were also well attended; we noticed decreased absenteeism on professional development days.

Farah and Barnett (2021) also encourage teachers to request time on professional development days to meet with English language instruction specialists or bilingual teachers to review data and discuss strategies to improve instruction for their multilingual students. Because it is often challenging to schedule collaborative team meetings with special teachers during the school day, professional development days are a great time to provide this opportunity.

Finally, teachers should set actionable goals. If teachers set actional goals for the school year, especially addressing the needs of the multilingual learners in their classrooms, they can request more targeted professional development sessions from their district. Student achievement is affected by teacher expectations of success for both the students and for themselves. What teachers expect students to learn influences student outcomes (Short, Becker, Cloud, Hellman, & Levine, 2018). A teacher with high expectations will exhibit positive behaviors toward students, motivating them to perform at a higher level because of their personal relationship (Short et al., 2018). Districts need to reevaluate their school-wide professional development plans and ensure that general education teachers are included in training that covers language acquisition and development—not just English language instruction—and that these teachers are supported beyond the training day. And teachers should do all they can to access professional development offerings that train them to support multilingual learners in their classrooms.

Teacher Preparation

In the introduction (page 1), you learned that Hispanic, African American, and Asian students in public K–12 classrooms have now surpassed the number of non-Hispanic Whites, and that this new majority will continue to grow (Maxwell, 2014b). Why is this such a concern? Because despite the growing number of multilingual learners in American schools, the teaching force's ethnic background has remained predominantly young, middle-class, non-Hispanic White females possessing little knowledge about these new diverse groups of students in their classes (Lin & Bates, 2014; Spiegelman, 2020).

Figure 1.3 illustrates the percentage distribution of teachers by race or ethnicity from 2017 to 2018.

This disparity indicates that few teachers possess cultural or linguistic backgrounds similar to their students and raises questions about whether teachers are prepared to work with students of differing racial and cultural backgrounds, which often leads to misplaced perceptions and lower expectations (de Haan, 2019). Because teachers will increasingly encounter a diverse range of learners, every teacher must have sufficient breadth and depth of knowledge and skills to meet the unique needs of all students, including those who struggle with

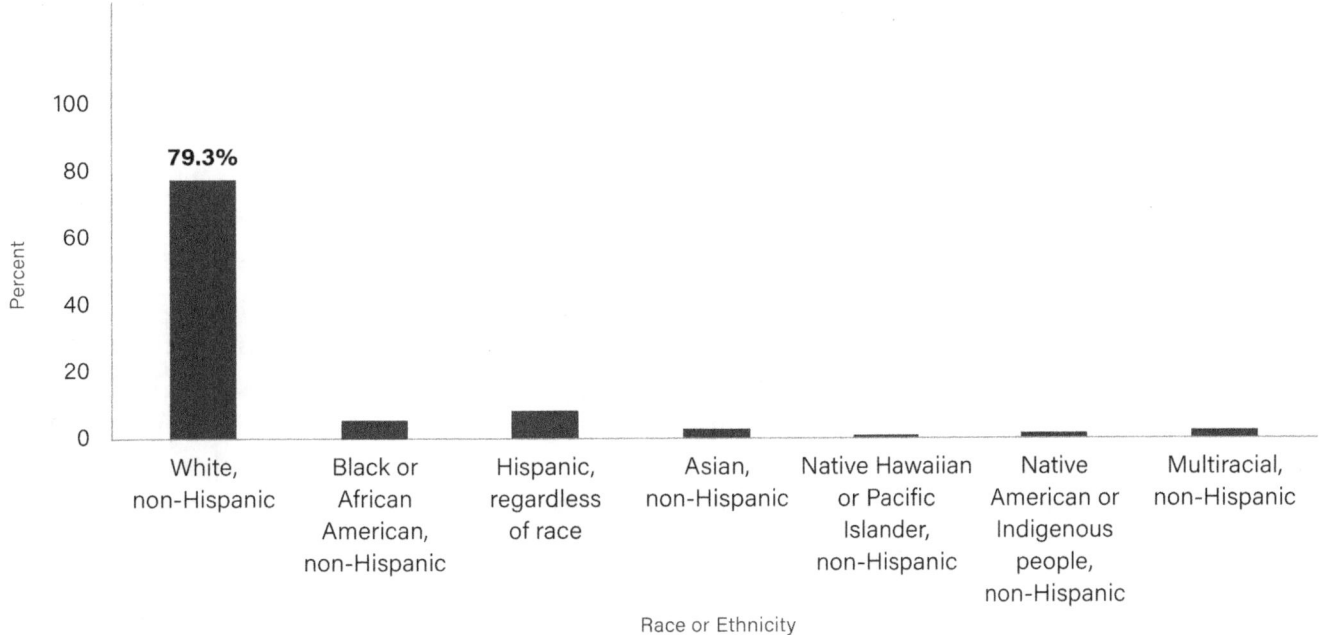

FIGURE 1.3: Percentage distribution of teachers by race or ethnicity from 2017 to 2018.

English (Samson & Collins, 2012). According to the U.S. Census Bureau, there are approximately 350 languages spoken in the United States, with most of the languages found in metro areas (U.S. General Services Administration, 2023). The top five non-English languages spoken in the United States are Spanish, Arabic, Chinese, Vietnamese, and Somali (Mitchell, 2020).

The changing classroom environment requires that we take a deeper look into general education teacher preparation programs that continue to fall short (Maxwell, 2014a) and the textbooks utilized in state teacher certification requirements. Research conducted on textbooks used in teacher education programs found that 3 percent or less contained content on multilingual learners (Hallman & Meineke, 2016; Watson, Miller, Driver, Rutledge, & McAllister, 2005). According to education researchers Jennifer F. Samson and Brian A. Collins (2012), there are inconsistencies across U.S. states in the required knowledge and skills regarding multilingual learners as a part of coursework. Furthermore, fifteen U.S. states have no requirements at all. Regarding teacher certification exams, many states do not specifically assess teacher knowledge and skills related to teaching multilingual learners. I would be remiss if I did not mention that there *are* teacher preparation programs that offer special certification tracks, such as teaching in urban schools, teaching in special education, and teaching multilingual learners. However, the majority of general education coursework does not require these specialized areas for degree completion nor an English language instruction certification, with the exception of a few states like California and Florida, where it is a required component of all teacher preparation programs (Buxton & Allexsaht-Snider, 2017).

This leaves content-area specialists, like STEM teachers, graduating without specialized coursework. But they are finding themselves responsible for teaching students without any formal training on how to teach the *content* and simultaneously teach *language*. This lack of training contributes to the low number of multilingual learners eligible for STEM, advanced mathematics, and advanced science classes (Zehr, 2011). A lack of formal training or coursework for teacher candidates can have other implications, including the lack of opportunities to reflect on their biases about multilingual learners' capabilities (Muñiz, 2019).

The United States must significantly improve educational outcomes for this growing and diverse majority of multilingual learners by addressing teacher programs that are graduating teachers who are ill-equipped to effectively teach culturally diverse students (Yuan, 2017).

Until changes are made to teacher education programs, there is something we can do to address this issue at the district level. We must provide teachers with embedded professional development on topics like understanding the language acquisition process and critical introspection of teacher beliefs about multilingual learners and linguistic diversity (de Haan, 2019). If we fail to do this, our education system will continue to imperil multilingual learners' academic success and emphasize what they cannot do rather than what they already do well, thereby stifling their learning because of low expectations (de Haan, 2019).

Consider the following proactive, actionable steps teachers can take when they recognize and accept that their preparation to teach multilingual learners was insufficient to enhance their skills and confidence.

- Seek professional development.
  - Look for workshops, seminars, and courses that address specific areas where you feel inadequately prepared.
  - Explore online platforms for valuable professional development resources.
- Take advantage of mentorship opportunities.
  - Ask your administrator about a mentor within the school or district or seek out one through your union or professional networks. Look for an educator who has more experience with multilingual learners and can provide guidance, support, and feedback. A mentor can help you navigate challenges and grow professionally.
- Join professional networks and communities.
  - Engage with professional associations related to education, such as the National Education Association or subject-specific groups. These networks can provide support, resources, and advice from experienced educators.
  - Participate in online forums and social media groups where educators share insights and strategies.
- Request peer observation.
  - Arrange through your building administrator to observe your peers' classes to gain insights into different teaching methods for multilingual

- learners. Also, invite trusted colleagues to observe your classes and provide constructive feedback.
- Engage in reflective practice.
  - Regularly reflect on your teaching practices and student outcomes. Keep a teaching journal to note what works and what doesn't and consider how to improve your methods.
- Enhance subject matter expertise.
  - Consider taking additional courses or studying to deepen your understanding if you feel weak in your subject area. Educational publishers and universities often offer content-specific resources and textbooks.
- Explore online resources.
  - Utilize the wealth of free educational resources available online. Websites like Khan Academy, TED-Ed, or specific educational YouTube channels can provide both content knowledge and pedagogical approaches.
- Request school support.
  - Discuss your needs with school administrators to seek support for further training or resources that could help you develop your skills.
- Attend conferences and workshops.
  - Educational conferences and workshops can offer new perspectives and tools that enhance your teaching. They also provide a chance to network with other educators and learn from their experiences.

By taking these steps, teachers can significantly enhance their capabilities and self-confidence, leading to more effective teaching and better outcomes for their multilingual students.

Early Grades Impact

Early grade, the period between infancy and third grade, is critical to developing STEM-related thinking dispositions, such as curiosity, inquiry, assessment, and analysis (Leung, 2023). Science is a student's first critical STEM exposure, and effective teaching in grade school is a make-or-break factor in future STEM success (Will, 2018). However, according to a study conducted by the National Council on Teacher Quality, only 3 percent of undergraduate elementary education programs require relevant coursework in biology, chemistry, physical science, or physics. Of the 810 programs studied, 66 percent do not require coursework in any of those core subjects (Will, 2018). Teachers need to act on and take advantage of young students' natural curiosity if they want STEM interest to increase in the middle grades.

We need to look at the science curriculum of our youngest learners, because that is where the STEM pipeline begins. Teachers and school leaders must reflect on their perception of their students because it impacts their teaching approach. Such reflection is necessary before school leaders will be able to create and sustain inclusive school communities for all students, especially multilingual learners (Cooper, 2021). Districts should provide teaching staff with ongoing, embedded professional development on how to successfully teach this growing population because school may be the multilingual learners' only opportunity for STEM exposure and increasing teacher self-efficacy can improve teacher retention in low-income communities.

Because children are innately curious, introducing STEM to students in preK through third grade will look quite different than introducing STEM during middle grades or high school. Students in preK through third-grade STEM classes can solve problems with hands-on exploratory skills. Consider the following example activities.

- Observe wool and compare it to human and other animal hair. Have students collaborate to find out why all hairs are not the same.

- View pictures of an animal's environment for students to see if the hair's texture might have something to do with the animal's environment and compare that to the different types of clothing used in different climates.

- Use dominoes to mimic how waves travel so students begin to understand that waves don't necessarily move, but push the next domino, that pushes the next domino. Students may then fill a clear bag with water, oil, and food coloring and gently rock the bag, watching the waves and making connections between these waves and the dominoes.

- Teach young students about mechanical energy by asking a simple question such as, "Why do you think some trucks and cars have big wheels and small wheels?" Provide simple materials like straws and cardboard disks and instruct students to create wheels and axles. Have students predict the distance traveled by the small-disk wheel compared to the large-disk wheel.

These are authentic, collaborative STEM activities that get students to think critically while working together like scientists to find solutions to problems. This makes learning fun and memorable. It introduces young students to STEM, encourages a love of learning, and promotes analytical skills.

Key Takeaways

STEM education offers tremendous benefits for every student, regardless of the type of career they may have in mind. A primary benefit of STEM education is its alignment with skills that are projected to be in high demand. Multilingual learners often exhibit a strong interest in STEM fields, yet structural barriers significantly impede their participation and success in these areas. Challenges such as language barriers that complicate students' understanding of complex technical vocabulary and concepts, a lack of culturally responsive teaching materials, and insufficient support services in schools can deter these students from fully engaging with STEM curriculum. Moreover, these learners might not have access to advanced STEM courses due to school and district practices that underestimate their capabilities or place undue emphasis on language proficiency over content understanding and cognitive abilities. Such educational practices not only prevent multilingual learners from pursuing their interests and talents in STEM, but also contribute to broader inequities within the workforce and higher education. Addressing these structural barriers is essential to foster inclusivity and equal opportunities in STEM education, thereby enabling multilingual learners to contribute their unique perspectives and skills to these critical fields that can also be instrumental in reversing the cycle of poverty.

Consider the following key takeaways from this chapter.

- Socioeconomic barriers significantly hinder the creation of a level playing field for multilingual learners, profoundly affecting their educational opportunities and outcomes.
- The concept of the leaky STEM pipeline refers to the observation that many students—particularly women and underrepresented minorities—start with an interest in STEM fields but drop out at various stages throughout their education due to various structural barriers, such as the inability to pay program registration fees, lack of prerequisite knowledge, competitive application processes, inability to demonstrate preexisting interest in science, poor literacy skills, lack of transportation, and a dearth of accessible opportunities.
- Increasing teacher self-efficacy is crucial because it directly impacts the quality of education that students receive. Teachers with high self-efficacy are more likely to implement innovative teaching strategies, manage classrooms effectively, and respond adaptively to students' needs.
- Effective teacher preparation for teaching multilingual learners is essential, as it equips educators with the specific strategies and knowledge needed to support this diverse student population. Proper training helps teachers understand language acquisition processes, cultural nuances, and the educational challenges that multilingual learners face.

CHAPTER 2

Understanding Multilingual Learners' Unique Needs

> *Languages carry the history and stories of a people.*
>
> — **DAVID CRYSTAL**

Understanding multilingual learners' unique needs requires recognizing the intricate interplay between language acquisition and academic achievement. These learners must navigate the dual challenges of mastering a new language while also engaging with grade-level content across various subjects. Effective support for multilingual learners involves providing targeted language instruction, scaffolding academic content, and creating a classroom environment that values linguistic diversity as an asset. Teachers can enhance learning by incorporating students' native languages and cultural contexts into lesson plans, which aids in comprehension and boosts confidence and participation. Acknowledging and addressing these needs ensures that multilingual learners are equipped to succeed academically and socially in their new educational settings.

Now that you understand the leaky STEM pipeline, its causes, and possible remedies, let us home in on multilingual learners themselves. Who are they? What are their specific needs? What common challenges do they face in education? And what unique strengths do they bring to the classroom? In this chapter, I define the term *multilingual learner* and share factors unique to their experience in the general education classroom, including socioeconomic status, culture shock, and cultural diversity. Given that students' educational experience is shaped by these factors, they have unique needs as learners. These include understanding language acquisition and proficiency levels, altered instructional design, additional time, and scaffolding. I end the chapter by discussing how STEM can meet these needs.

Multilingual Learners' Assets

If you were to look up *multilingual learner*, you would find a plethora of definitions. However, there are four that are relevant to this book's context.

1. Persons whose primary, native language is a language other than English, and represent a population of nearly five million students, or 10 percent of students enrolled in K–12 public schools (National Center for Education Statistics, 2022).

2. Students who demonstrate a sufficient difficulty in reading, writing, speaking, or understanding the English language—a challenge that "inhibits their ability to learn successfully in classrooms where English is the language of instruction or to participate fully in the larger US society" (Language Magazine, 2018).

3. A student whose primary language is not English, and whose English proficiency, or lack thereof, provides a barrier to successful learning (Council for Exceptional Children, n.d.).

4. Language-minority students whose English proficiency affects their ability to meaningfully participate and succeed in school (Linquanti, Cook, Bailey, & MacDonald, 2016).

Definitions two, three, and four contribute to the practice of keeping culturally and linguistically diverse students out of rigorous classes. Those three definitions reflect deficit-based views of multilingual students. When deficit thinking is applied in classroom settings, the results often include segregation of students who are viewed as inferior and arguments about the educability of certain groups of students that rely on pseudoscientific beliefs in cultural or genetic deficits (Wang et al., 2021). While not being able to speak, read, or write in English does impact learning, it does *not* inhibit or act as a barrier to a student's ability to learn successfully. The first definition is a more proper and accurate definition of multilingual learners that general education teachers should adopt because it defines learners based on facts and does not draw false connections between language proficiency and cognitive ability.

Recognizing and celebrating multilingual learners' assets involves acknowledging the rich linguistic and cultural backgrounds they bring to the classroom. Educators and peers can foster an inclusive environment by viewing these diverse language skills not as hurdles to overcome but as valuable resources that enhance learning for everyone. To do this effectively, teachers can integrate multilingual resources into their teaching materials and encourage students to use their home languages in class discussions and group projects. This not only validates the students' cultural identities but also promotes a deeper understanding among all students.

Viewing multilingual learners through this asset-oriented mindset allows teachers to remember that each student has strengths that support their learning and contribute to the classroom community. Multilingual learners also experience challenges outside of education after moving to an English-speaking country. The following section explores some of these factors.

Factors Unique to Their Experience

This section explores the challenges multilingual learners face when they move to an English-speaking country, with a particular focus on the phenomenon of culture shock. As individuals from diverse backgrounds transition into a new cultural environment, they often encounter unexpected differences in social norms, language, and daily life practices. This adjustment period can be both disorienting and enlightening, offering unique obstacles and opportunities for personal growth. Understanding culture shock is crucial for educators and community members who support these learners in navigating their new world.

Culture Shock

Culture shock is a term used to describe what happens to people when they encounter unfamiliar surroundings and conditions, such as being transplanted into a foreign setting (Simon Fraser University, n.d.). This is relevant for teachers of multilingual learners, as many of them enter the general education classroom after moving from another country. Students experience culture shock in the following four stages.

1. **Honeymoon stage:** The learner is very positive and curious, anticipating new and exciting experiences. They even idealize the host culture.

2. **Frustration stage:** The learner starts to feel that what is different is inferior. The host culture is confusing, or the systems are frustrating. The loss of control and lack of understanding can cause a range of emotions like anxiety, depression, and fear of the unknown. Students may overreact to seemingly small irritations, like misplacing an item or spilling a drink. At this stage, they may blame their frustrations on the new culture (and its shortcomings) rather than on the adaptation process.

3. **Adjustment stage:** The learner feels more relaxed and develops a more balanced, objective view of their experience.

4. **Acceptance stage:** The learner feels a new sense of belonging and sensitivity to the host culture.

Note that if students adjust to a new culture and then return to their country of origin, they may experience *reentry shock* as they readjust to cultural differences because culture shock is cyclical rather than linear in nature. A student may enter at any one of the four stages, and even if they exit that stage, they may reenter it later. Figure 2.1 (page 28) illustrates the four stages of culture shock (Simon Fraser University, n.d.).

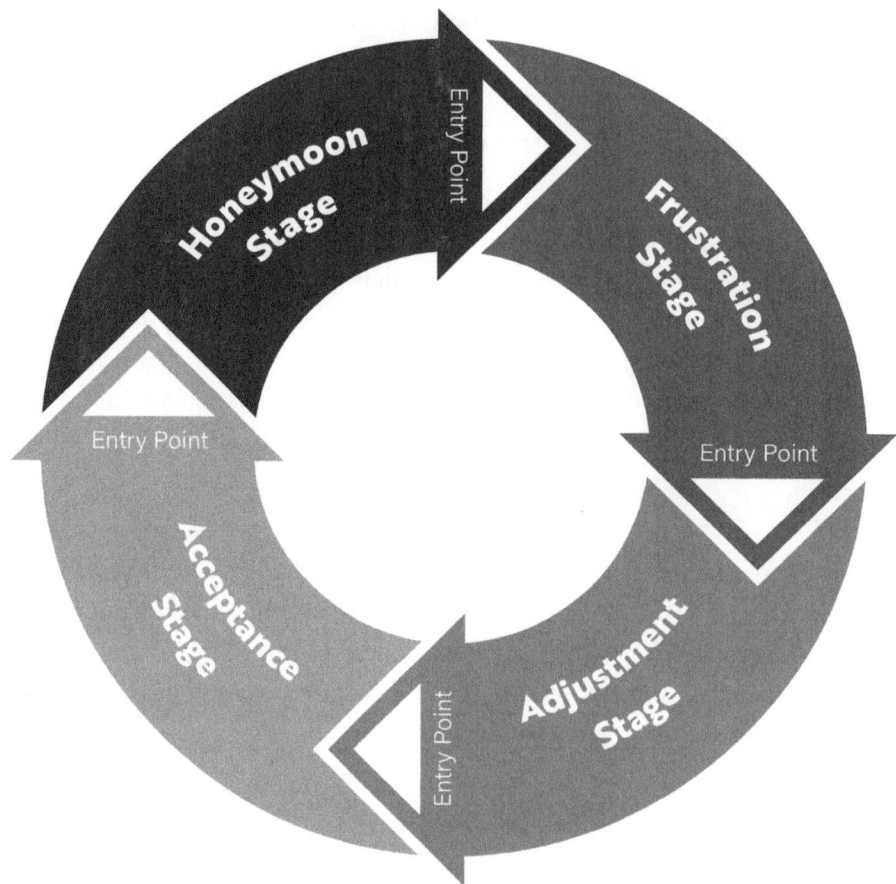

FIGURE 2.1: The cycle of culture shock.

Cultural Diversity

Multilingual learners bring many different assets to the classroom, including linguistic, academic, and social-emotional needs. They also bring information and ideas they've already learned, as well as home and community experiences that can be used for standards-based instruction. Multilingual learners also have their own level of academic achievement, and educators can help them see how they can use their background competencies to learn in their current academic environment.

The rich tapestry of cultural identities that multilingual learners bring to public schools significantly shape their educational experiences and interactions. The values, traditions, and perspectives from their countries of origin can contrast markedly with cultural norms of English-speaking countries and educational practices, such as educational expectations, communication styles, and teacher-student relationships. For example, students may be used to a more formal classroom atmosphere where direct questioning of teachers is discouraged, whereas U.S. classrooms often promote a more interactive and open dialogue between students and teachers. This disparity can lead to misunderstandings or a sense of alienation among multilingual learners as they navigate these new social and academic expectations.

Additionally, multilingual learners might face challenges, such as language barriers, stereotyping, or even bullying, which can affect their self-esteem and academic performance. Schools that celebrate diversity and implement inclusive practices can significantly enhance these students' experiences. Programs like bilingual education, multicultural festivals, and support groups can help bridge cultural gaps and promote a more welcoming environment. Such initiatives not only support the academic success of multilingual learners, but also enrich the educational experience for all students by fostering a deeper understanding and appreciation of global cultures.

Language Acquisition and Proficiency

Multilingual learners come from a variety of cultural, educational, and linguistic backgrounds, and don't all fit neatly into a single label. Multilingual learners have different proficiency levels and may have varying needs that affect their ability to interact in the classroom. For example, multilingual learners may need additional time for oral language practice to build proficiency. Given that they must simultaneously attain English proficiency and master the required grade-level content knowledge, it is essential for educators to grasp the fundamental principles of second language acquisition. A lack of understanding among teachers and administrators regarding this process can lead to several negative outcomes, including the following.

- Misguided referrals to special education services
- Communication barriers due to reliance on a language that multilingual learners may not understand yet
- Insufficient support for multilingual learners in acquiring new content knowledge in English

It is important to always remember that second language proficiency develops gradually and differs among individuals. By recognizing the specific stage of English-language proficiency that a student has reached, teachers can tailor their instructional strategies to effectively meet the student's linguistic needs.

Table 2.1 (page 30) illustrates the stages of language acquisition.

The chart provides explanations of what students can do at each language-acquisition stage. By aligning instructional strategies with the specific stages of language acquisition, teachers can more effectively support multilingual learners in their journey to fluency. Each stage requires different types of support, and recognizing where students are in their language development is key to providing the right kind of instruction to foster both confidence and competence.

Language acquisition and language proficiency are closely intertwined concepts that describe the levels at which individuals learn and use a second language. *Language acquisition* refers to the process through which learners gain knowledge of a new language, including listening, speaking, reading, and writing skills. Language proficiency, on the other hand, measures how well an individual uses the language across various domains of speaking, listening, reading, and writing.

TABLE 2.1: Language Acquisition Stages

Stage	Student Abilities
Preproduction	This is also called *the silent period*, when the student takes in the new language but does not speak it. This period often lasts six weeks or longer, depending on the individual.
Early Production	The individual begins to speak using short words and sentences, but the emphasis is still on listening and absorbing the new language. There will be many errors in the early production stage.
Speech Emergence	Speech becomes more frequent, words and sentences are longer, but the individual still relies heavily on context clues and familiar topics. Vocabulary continues to increase and errors begin to decrease, especially in common or repeated interactions.
Beginning Fluency	Speech is fairly fluent in social situations with minimal errors. New contexts and academic language are challenging, and the individual will struggle to express themselves due to gaps in vocabulary and appropriate phrases.
Intermediate Fluency	Communicating in the second language is fluent, especially in social language situations. The individual can speak almost fluently in new situations or in academic areas, but there will be gaps in vocabulary knowledge and some unknown expressions. There are very few errors, and the individual demonstrates higher-order thinking skills in the second language, such as offering an opinion or analyzing a problem.
Advanced Fluency	The individual communicates fluently in all contexts and can maneuver successfully in new contexts and when exposed to new academic information. At this stage, the individual may still have an accent and use idiomatic expressions incorrectly at times, but the individual is essentially fluent and comfortable communicating in the second language.

Source: Adapted from Robertson & Ford, n.d. Used with permission.

It is often assessed through standardized tests that evaluate a learner's ability to communicate effectively in the target language. Proficiency levels help educators tailor their instruction to meet their students' specific linguistic needs at different language development stages.

The World-Class Instructional Design and Assessment (WIDA) is an essential resource that supports the education of multilingual learners primarily in the United States (WIDA, n.d.b). It provides a framework and tools for assessing English language proficiency and developing appropriate instructional strategies. WIDA's standards outline key language skills that multilingual learners need to succeed in academic settings, integrating language development with content learning. The WIDA assessments, such as the ACCESS for ELLs, are designed to measure students' academic language proficiency in English, which aids educators in understanding the progression of language acquisition among students. This framework supports consistent assessment and reporting across states and guides teachers in creating effective learning environments that promote both language acquisition and content-knowledge mastery (WIDA, 2020).

WIDA is used by a group of states to measure students' progress in English language development. This comprehensive system is used by members of the WIDA Consortium, a U.S.-based collaborative group of forty-two member states, territories, and federal agencies. Visit https://wida.wisc.edu/about/consortium to view a map of WIDA consortium members.

Table 2.2 (page 32) displays WIDA's six levels of language proficiency.

The six levels of language proficiency shown in table 2.2 describe the student's performance in the following four domains: reading, writing, listening, and speaking. Knowledge of a student's six proficiency levels can greatly assist teachers in tailoring instruction to meet multilingual learners' specific linguistic needs. These levels range from entering (level 1) to reaching (level 6), and each level provides a detailed description of the language skills multilingual learners are typically able to demonstrate in the four domains (Shenandoah County Public Schools, n.d.).

Altered Instructional Design: Gamification

I want to discuss the video game model and why it should be considered in your lesson planning. I am not speaking of using actual video games in your instruction. I am speaking of *gamification*—using video game *mechanics* in your instructional design. You may be wondering how gamification fits into a chapter about understanding multilingual learners' needs. It's a perfect fit because the video game model can be highly effective in supporting multilingual learners by leveraging the engaging and interactive nature of gaming and using students' everyday experiences, interests, and knowledge to support their learning. This model incorporates elements of gameplay—such as immediate feedback, progressive challenges, and interactive storytelling—into the learning process (Dehghanzadeh, Fardanesh, Hatami, Talaee, & Noroozi, 2019).

Although it sounds like a strategy to solely make learning more engaging, it is also a student motivation tool. In games, progress encourages progress. This aspect, when brought into the classroom, can encourage students to be more self-motivated and incentivized in their own learning. This is commonly done using badges, leaderboards, point systems, and levels. Not only will it benefit your multilingual learners, but it will also benefit all your students and help them develop perseverance and a growth mindset.

If used correctly, gamification can do the following.

- Make students feel more empowered and confident
- Motivate students to continue trying and keep practicing
- Help build a better classroom environment
- Help develop behavior skills

But it is also important to realize that with games comes competition, winning, and losing. So, make sure that your classroom is a safe and positive environment when implementing gamification. You can do this by defining the rules, processes, and procedures and

TABLE 2.2: WIDA's Six Levels of Language Proficiency

Level 1: Entering	Level 2: Beginning	Level 3: Developing	Level 4: Expanding	Level 5: Bridging	Level 6: Reaching
• Represents content-area language graphically • Uses words, phrases, or chunks of language when presented with commands, directions, or questions • May speak with phonological, syntactic, or semantic errors when presented with basic oral commands or direction	• Uses general language related to the content areas • Uses phrases or short sentences • May speak or write with phonological, syntactic, or semantic errors when presented with multi-step commands, directions, or questions	• Uses general and some specific language of the content areas • Expands on sentences when speaking or writing paragraphs • May have phonological, syntactic, or semantic errors when presented with oral or written narrative or expository descriptions	• Uses specific and some technical language of the content areas • Uses a variety of sentence lengths of varying linguistic complexity when speaking or writing paragraphs • Has minimal phonological, syntactic, or semantic errors when speaking or writing	• Uses specialized or technical language of the content areas • Uses a variety of sentence lengths of varying linguistic complexity when engaging in oral discourse or writing stories, essays, or reports • Speaks and writes comparable to that of English-proficient peers when presented with grade-level material.	• Uses specialized or technical language reflective of the content areas at grade level • Uses a variety of sentence lengths of varying linguistic complexity in extended oral or written discourse as required by the specified grade level • Speaks and writes in English comparable to English-proficient peers

balancing the reward of success with the pressure to succeed. Keeping in mind the following elements of gamification can help students avoid seeing losing as failure (Wanasek, 2023).

Elements of gamification include the following (Alimansyah, 2023; Castelán, 2023; Wanasek, 2023).

- **Rewards:** *Rewards* are incentives given for achieving specific goals or challenges. They reinforce positive behavior and motivate continued engagement.
- **Leaderboards:** *Leaderboards* rank users based on their performance relative to their peers. They foster competition and can motivate participants to improve their standing by engaging more deeply with the activity. Start small by awarding students points based on their class participation. If your classroom isn't ready for the competition of a leaderboard, students can be assigned codenames to keep their identities hidden.
- **Challenges:** *Challenges* are specific tasks that users must complete to earn rewards or recognition. They add variety and excitement to the activity, encouraging users to engage in behaviors that will help them achieve their goals.
- **Feedback:** *Feedback* helps users understand how well they are doing and what they need to improve; immediate and ongoing feedback is crucial in gamification. Feedback mechanisms can include visual, auditory, or textual responses to user actions.
- **Avatars:** *Avatars* represent users in the game or gamified system. They can be customized and developed over time, which can increase the user's emotional investment in the activity.
- **Narratives:** A compelling *narrative* can provide context and reasons for users to care about their actions within the gamified system. Integrating storytelling into the experience can greatly enhance engagement.

According to *Gaming the Classroom: The Art and Science of Game Based Learning* here are some stats to get you thinking about why gamification should be considered in education (eLearning Infographics, 2018).

- Gamers spend an average of thirteen hours weekly playing.
- Ninety-seven percent of youths play computer and video games.
- There are 1.8 billion gamers worldwide.
- There are an estimated 194 million gamers in the U.S.
- Eighty percent of learners would be more productive if learning were more game-like.

In education, we discuss growth mindset versus fixed mindset at length and how we want students to have a growth mindset, especially in mathematics and science (Dweck, 2007). As educators, we often hear students say they don't have a mathematics brain or a science brain and watch them give up once they begin to struggle. Imagine how this negative mindset affects a student who must learn content in a language that is not their native language. The negative mindset puts students' brains in a state of stress.

Stress, which can include boredom and lack of understanding, causes the brain's alert system (the amygdala) to go into high alert and shut down all communication with the thinking brain (prefrontal cortex). A brain in stress mode goes directly to the involuntary actions of fight, flight, or freeze. When in this stress-response state, very little learning occurs, and the student really cannot control their behavior because the involuntary brain is in charge. This must occur for survival. It is important to understand that the brain seeks two main things: pleasure and survival. The brain gets pleasure from the release of the neurotransmitter dopamine, and it survives due to the involuntary actions of the brain during fight, flight, freeze, or fawn (Desautels, 2012).

So, why do gamers—who experience over 80 percent repeated failures, setbacks, and increasingly challenging work—continue to persevere when playing video games? Why do they *not* create a negative fixed mindset of failure? It has to do with motivation. Neurologist and teacher Judy Willis (2011) explains that motivation from positive experiences is necessary to overcome the brain's resistance to applying effort when there is a low expectation of reward or success. Positive experiences release dopamine, a neurotransmitter that gives the sensation of pleasure, and video games are designed to release dopamine by always including four factors (Willis, 2011).

1. **Goal clarity:** Video games always have a goal—save the world from zombies, for example.

2. **Achievable challenges:** Achievable challenges refer to tasks or goals that are suitably demanding for the learner—neither too easy nor too difficult—but are realistically attainable with effort.

3. **Predictions:** A big part of video games is players having to guess what their next move should be.

4. **Feedback:** The other reason video games are so successful is because the students receive immediate feedback. With video games, they don't have to wait to find out if their prediction was correct or not. They will either lose energy or a life, or the game will give them a loud sound letting them know if they were correct or incorrect.

These same four factors, if included in your lesson design, can aid your English learners in finding success in STEM content areas.

1. **Goal clarity:** The students know what they must accomplish and what the expectations are. When working with English learners, start the lesson by selecting the smallest amount of material that will have the maximum meaning for the learner. Know the prerequisite skills required for the task and help the students gain this knowledge by using visuals like videos or pictures, and use the students' funds of knowledge to help them understand and get the maximum understanding of the lesson's goal—what it is they must accomplish.

2. **Achievable challenges:** Finding a students' achievable challenge is the same as setting levels for the students to work at. You can determine the students' level of achievable challenge by considering what is achievable for them based on their funds of knowledge and level of language ability. Differentiate the instruction and requirements based on these determinations. Consider starting the lesson where most students have similar levels of knowledge or create intentional groups where everyone can contribute. See chapter 4 (page 77) where I explain the intentional groups that I call Two Pairs in a Quad.

3. **Predictions:** In the classroom, this can occur by making sure that the practice occurs first in your presence over a short period of time while the student is focused on the learning. For example, if the student is working on a STEM task where they must decide what material to use in building a house that stays cool, have the student predict which material they think will be the best. In your presence, have the student follow the process you modeled to see the outcome. Explain that they must do the same to the other materials to determine whether their prediction was correct or incorrect. Use sentence frames (chapter 6, page 155) to help students formulate sentences to describe their prediction and write their results.

4. **Feedback:** In the classroom, we need to provide the same kind of immediate feedback as found in video games. Using our STEM task of the house as an example, you would need to watch the student practice the procedure on their selected material and provide them with prompt and specific feedback on what variable needs to be altered to correct or enhance the outcome.

So, imagine how positive experiences can impact a multilingual learner's success in a STEM discipline while learning English. A positive mindset keeps the brain out of stress and therefore avoids fight, flight, or freeze, and the brain is open to receive new information to the prefrontal cortex.

Students' successes and their improved confidence and attitudes are benefits of the video game model. The immediate feedback on their prediction releases dopamine, which makes them happy and makes them want to continue trying in order to get the good feeling that dopamine provides. Just think about yourself: the video game model is everywhere in your life in the form of credit card, hotel chain, and airline rewards. All these rewards—points earned and perks of free rooms, flights, or other amenities—are a part of the video game model. So, if you can use this model with your English learners while modifying it to the students' level of language acquisition, imagine the confidence and success they will experience in the STEM classroom!

Attention Span

There are two key aspects of timing to consider when meeting multilingual learners' unique needs. First is the length of time a teacher talks during direct instruction. Merriam-Webster defines *attention span* as the length of time a person or group can concentrate or maintain interest (Attention Span, n.d.). Knowing the attention span per students' age is paramount to understanding how long your direct instruction should last before either checking for understanding (formative assessment) or having the students do an activity connected to what you discussed to solidify the learning (Brain Balance, n.d.). This keeps the brain from going into fight, flight, freeze, or fawn and keeps the amygdala out of stress mode. Table 2.3 shows attention span by age.

TABLE 2.3: Average Attention Span By Age

Age range	Attention span in minutes
Two years old	Four to six minutes
Four years old	Eight to twelve minutes
Six years old	Twelve to eighteen minutes
Eight years old	Sixteen to twenty-four minutes
Ten years old	Twenty to thirty minutes
Twelve years old	Twenty-four to thirty-six minutes
Fourteen years old	Twenty-eight to forty-two minutes
Sixteen years old	Thirty-two to forty-eight minutes

Teachers can create a more supportive and effective learning environment for multilingual learners by tailoring lessons to students' natural attention spans, which can significantly enhance the effectiveness of instruction. Understanding that students generally have short attention spans depending on their age, developmental stage, and individual differences helps teachers design lessons that maintain engagement and maximize learning. Using this information will help mitigate the potential frustration associated with language barriers and contribute to a more inclusive and engaging classroom experience.

Second, students benefit from a pause between a teacher posing questions and requesting student answers. Offer students silent think time before cueing them for a response. Dylan Wiliam—consultant, educator, and emeritus professor of educational assessment at the UCL Institute of Education—uses a chant to remind teachers to provide more wait time when asking for a students' reply to a question, "One, two, three, four got to wait a little more" (Hamid Mahmood, 2016). This chant provides students with more wait time, and helps you control your desire to intervene too quickly. You can say it as often as you like if you want to provide additional wait time. This provides multilingual learners with the following benefits.

- **Processing time:** Multilingual learners need additional time to process information due to their language acquisition process. Providing think time allows them to comprehend the content and formulate responses effectively.

- **Language production:** Multilingual learners require extra time to translate their thoughts into the target language. Think time supports them in articulating their ideas and expressing themselves more confidently.

- **Cognitive processing:** Multilingual learners may need time to access prior knowledge and connect it to new concepts. Think time facilitates cognitive processing and enhances their understanding of the content.
- **Equity and inclusion:** Offering think time ensures that multilingual learners have an equal opportunity to participate and contribute to discussions, promoting inclusivity in the classroom.

Scaffolded Instruction

Scaffolding is not just another word for help. Scaffolding is a special word for help that assists learners in moving toward new skills, concepts, and levels of understanding (Gibbons, 2015). *Scaffolding* is an instructional technique wherein educators provide successive levels of support to help students reach higher levels of comprehension and skill acquisition. For multilingual learners, scaffolding plays a crucial role in facilitating language and content acquisition for several reasons.

- **Bridges prior knowledge and new information:** Scaffolding links what English learners already know in their native language to new English concepts, making learning more relatable and digestible.
- **Allows gradual release of responsibility:** Starting with teacher-led demonstrations or explanations, scaffolding gradually transitions the responsibility of learning to the student. As English learners become more proficient, less support is provided, fostering their autonomy and confidence.
- **Contextualizes language:** Scaffolding often uses visuals, realia (everyday objects), gestures, and other contextual clues. This helps English learners understand the meaning behind new vocabulary and concepts, even if they aren't familiar with every word.
- **Reduces cognitive load:** By breaking down information into manageable chunks and providing support at each stage, scaffolding prevents English learners from becoming overwhelmed, allowing them to focus on specific aspects of the language or content.
- **Supports skill development:** Scaffolding not only aids in language acquisition, but also supports the development of academic skills. For instance, a scaffolded approach to reading might start with identifying key vocabulary, then summarizing paragraphs, and finally analyzing the entire text.
- **Promotes interaction:** Many scaffolding techniques, like pair work or group discussions, encourage multilingual learners to interact with peers. This provides authentic opportunities for language practice and reinforces learning through collaboration.
- **Encourages meaningful participation:** With the right scaffolds in place, English learners can participate in discussions, projects, and other classroom activities that they might otherwise find too challenging. This boosts their engagement and sense of belonging.

- **Provides continuous feedback:** Scaffolding inherently involves regular check-ins and adjustments based on the learner's progress. This continuous feedback helps English learners understand their strengths and areas for improvement, guiding their learning journey.

In summary, scaffolding is instrumental in guiding English learners through the intricate process of language acquisition, ensuring they have the tools and support necessary to understand, engage with, and eventually master both the English language and the content being taught.

STEM as a Solution

STEM has a direct impact on language development for multilingual learners because of its inquiry-based, hands-on instructional practice. During STEM classes, multilingual learners are provided with the opportunity to practice the language by working in collaborative small groups. Language development requires active learning, and STEM encourages listening and speaking, which are necessary in the acquisition of listening and speaking skills (Short et al., 2018). Research shows that multilingual learners can learn STEM content and practices while simultaneously building proficiency in English when they are allowed to interact with content in varied ways, build on what they already know, and develop new technical knowledge. In other words, language development and concept development can occur simultaneously (Francis & Stephens, 2018; Roesler, 2022).

It is imperative that all educators understand that multilingual learners, at all levels of language proficiency, can be successful in STEM content classes and higher mathematics and science classes when the instruction is engaging, meaningful, interactive, and respects and appreciates the assets that multilingual learners bring to the classroom (Francis & Stephens, 2018). It is also very important to remember that some immigrant English learners have strong academic backgrounds. Some are above equivalent grade levels in certain subjects—mathematics and science, for example. There are many multilingual learners who are literate in their native language and may have already studied a second language. Much of what these learners need is English language development so that they become more proficient in the language and transfer the knowledge they learned in their native country's schools to the courses they are taking in English (Echevarria, Vogt, & Short, 2012). But all multilingual learners, regardless of their level of language ability, need opportunities to interact with the language to develop their academic vocabulary. However, if the general education teachers do not have the training, confidence, or knowledge on how to teach language and content simultaneously, multilingual learners will continue to drop from the leaky STEM pipeline and be excluded from the exposure to a career that could positively alter a person's socioeconomic status.

Key Takeaways

Incorporating gamification and scaffolding and understanding student attention spans are important components of effective instructional strategies for all students, but especially multilingual learners. Using gamification strategies for multilingual learners is a powerful tool for enhancing engagement and motivation and in making learning more interactive, accessible, and enjoyable. It also promotes active participation and provides immediate feedback, helping students to stay motivated and track their progress. Scaffolding is essential for supporting multilingual learners effectively by providing support at each step until students become more proficient with the content. Scaffolding allows multilingual learners to build on their existing knowledge and gradually develop new skills. Additionally, by incorporating attention span limits, educators can reduce pressure on multilingual learners and provide the processing time they need to improve their comprehension. Together, these approaches create a supportive and inclusive learning environment where multilingual learners can thrive.

Consider the following key takeaways from this chapter.

- Incorporating gamification into students' educational experiences while learning in a nonnative language offers a powerful tool to enhance their learning in multiple dimensions, making the process more accessible, enjoyable, and effective.

- Incorporating achievable challenges into educational practices involves understanding each student's current mastery level and crafting lessons that advance that mastery in incremental, engaging steps. This strategy supports academic success and fosters a love of learning and resilience in students—key traits for lifelong learning.

- By strategically leveraging the WIDA proficiency levels, teachers can provide a structured, supportive, and effective learning environment that facilitates multilingual learners' language development and academic success. This approach ensures that instruction is responsive to learners' diverse needs at all stages of English language acquisition.

- By tailoring lessons to students' natural attention spans, teachers can create a more supportive and effective learning environment for multilingual learners. These strategies help mitigate potential frustration associated with language barriers and contribute to a more inclusive and engaging classroom experience.

- Second language proficiency develops gradually and differs among individuals. By recognizing the specific stage of English language proficiency that a student has reached, teachers can tailor their instructional strategies to effectively meet the student's linguistic needs.

- Multilingual learners' identities are not static; they evolve as students interact with their new environment. This dynamic process of identity formation can create a complex interplay between maintaining one's cultural heritage and adapting to new cultural norms. Educators play a crucial role in this aspect by providing support that respects and values the students' original cultures while simultaneously facilitating their integration into the school community. This delicate balance can empower multilingual learners to navigate their cultural identities confidently and contribute their perspectives to the learning environment, enriching the educational experience for everyone involved.

CHAPTER 3

Making Instruction Applicable Through Culturally Responsive Teaching

> *Culturally responsive teaching is not about making assumptions based on students' cultural backgrounds, but about empowering them by honoring their experiences, perspectives, and voices in the classroom.*
>
> **—ZARETTA HAMMOND**

One cannot begin a conversation about the lack of diversity in STEM classrooms without discussing culturally responsive teaching, which is deeply ingrained in brain science and directly linked to student engagement and improved classroom environment (Regional Educational Laboratory, 2019). Culturally responsive teaching is an asset-based pedagogy that incorporates students' cultural identities and lived experiences into classroom teaching as a tool for effective instruction (Will & Najarro, 2022). Research shows that teachers in public schools have historically undervalued the potential for academic success among students of color and have set low expectations for this population by seeing cultural differences as barriers rather than as an asset to learning and students' achievement (Will & Najarro, 2022). *Culturally responsive teaching* is a rich, intentional approach woven into every aspect of student learning that focuses on the assets students bring to the classroom rather than on their deficits (Understood.org, n.d.). This approach raises academic expectations and makes learning relevant for all students because the skills and academic knowledge are situated within the students' lived experiences and makes the learning more meaningful and thoroughly appealing (Understood.org, n.d.). According to writer and educator Zaretta

Hammond (2015), culturally responsive teaching is a pedagogical approach firmly rooted in learning theory and cognitive science that helps students build intellective capacity—the increased power the brain creates to process complex information more effectively and culture plays a critical role in this process.

Maintaining high expectations for all students begins at the institutional level. Districts must believe, embrace, and reform their way of looking at culturally different students and refrain from seeing this population of students through a deficit lens, but from a lens that sees each student's cultural identity, language, and family structure as a contribution to learning. Cultural competence connects to the personal dimension because both students and teachers need to understand their own cultural identity, biases, prejudices, and experiences that can impact the formation of meaningful connections or relationships with people from various cultural backgrounds. And finally, guiding students to develop critical cultural consciousness connects to the instructional dimension where theory meets practice. It is the place where the teacher relates the coursework to the students' culture, language, perspectives, and life experiences and no longer sees cultural differences as barriers but rather as assets of learning (Will & Najarro, 2022).

Culturally responsive teaching calls teachers to recognize the importance of including students' cultures in education, embracing multiple perspectives in the classroom, and using culture in curriculum. This gives all students the best chance of becoming academically successful, and they can feel like contributors to the classroom environment, knowing that their culture matters.

The U.S. Department of State's Bureau of Educational and Cultural Affairs offers multiple free courses on global education, global competence, as well as STEM innovations. The courses teach participants how cultures are instrumental in preparing students to be successful in a rapidly changing world. They define *globally competent students* as those who can combine disciplinary content knowledge by asking critical questions, analyzing multiple perspectives, and solving problems (U.S. Department of State Bureau of Educational and Cultural Affairs, n.d.). For this endeavor to succeed, teachers in the United States public education system must become culturally competent educators.

Educators today must be prepared to teach their students to succeed in a globalized tomorrow. The U.S. Department of State Bureau of Educational and Cultural Affairs (n.d.) writes that the interconnectedness of our world and the rise of new technologies demand that students be prepared to work across borders and engage with multiple perspectives in the pursuit of knowledge. The Center for American Progress notes that local communities must measure and be held accountable for instilling the dynamic set of skills and abilities that students will need to secure good jobs in the future (Jimenez, 2020). Developing these skills will be key to their success. But for that to occur, schools need to prepare *all* students for the jobs of tomorrow because students will need a set of specific skills that will enable them to fully participate, solve problems, and adapt in our changing world (Jimenez, 2020) because nearly every aspect of how Americans work has changed since the 1970s (Lisa, 2019).

Computer literacy skills; critical academic, technical, or crosscutting skills; and 21st century skills, such as critical thinking and collaboration, will all allow students to participate in the evolving workforce (Jimenez, 2020). Black people are overrepresented in support roles—such as food service, truck driving, and clerical roles—that are most often affected by advances in technology (Cook, Pinder, Stewart, Uchegbu, & Wright, 2019). Across three cities—including Gary, Indiana; Columbia, South Carolina; and Long Beach, California—Latinos are sometimes at even higher risk of job loss due to automation (Leins, 2019). But how can educators prepare students for this changing world if they themselves do not possess the tools, skills, or knowledge to be culturally responsive?

This lack of cultural responsiveness is evident in those districts that restrict multilingual learners from participating in STEM or advanced mathematics and science courses by placing a language proficiency level requirement for admittance, as if language ability is directly correlated to cognitive ability (Francis & Stephens, 2018). Restricting students from STEM participation cheats all students of these diverse perspectives where all cultures and identities are embraced.

Global competence in STEM requires students to develop and apply knowledge in culturally appropriate, relevant, and sustainable ways. For this to occur, educators must be equipped to understand the valuable role they play in developing cultural awareness in the curriculum, classrooms, and relationships. They must also recognize the grave global impact if public education fails to accomplish this.

When I first heard about culturally responsive teaching, my first thought as a mathematics and science teacher was that I had to find connections between the content and each student's culture. I immediately became overwhelmed and stressed at that thought. How could I possibly do that? Where would I look? How much time would that take? How would I keep up with the pacing guide set by my district?

I saw culturally responsive teaching as a humongous task that would be a burden to undertake. I didn't know what a culturally responsive lesson looked like, so how could I possibly succeed? I began to research culturally responsive teaching as I realized how little I understood about it and what it would require of me.

Scan the QR code to view my YouTube series on culturally responsive teaching.

In this chapter, I share with you what I learned through my research, starting with how to develop your cultural knowledge based on individualist and collectivist cultural archetypes. I discuss what it looks like for teachers to value their students' diverse cultures, which fosters strong teacher–student relationships and helps minimize misunderstandings. We, as educators, must identify our own implicit bias and how it might be influencing our relationships with students. With the help of fictional student Maria Christina Torres, I offer strategies you can use to incorporate culturally responsive teaching into your instruction. Toward the end of the chapter, I introduce the STEM challenge and show you how to plan and execute it, incorporating what you've learned about culturally responsive practices.

Culturally Responsive Teaching
youtu.be/HvrOZBHhyBw

Culture Informs Relationships

Although it is impossible to research the culture of every multilingual learner you encounter in your classroom, it's helpful to think in terms of patterns across cultures, what Zaretta Hammond (2015) calls "cultural archetypes" (p. 25). *Cultural archetypes* state that there are common values, worldviews, and practices among like cultures; though there may be diverse cultures represented in your classroom, there are patterns that unite them. One of the key ways to reduce confusion about how to attend to all your students' different cultures is to first identify which cultural archetype dominates: individualism or collectivism.

Dutch sociologist Geert Hofstede created a chart that represents one of the seven dimensions of the Cultural Dimensions Index. The chart lists countries according to a one-hundred-point scale in seven dimensions; one dimension is the level of individualism and collectivism within a society. A high number signifies an extremely individualist culture (such as the United States, Australia, and the Netherlands, which score in the nineties and eighties, respectively) whereas a low number signifies a more collectivist culture (such as Panama, Thailand, and Zambia, which score in the tens and twenties, respectively). Visit **go.Solution Tree.com/EL** for a link to view Hofstede's chart. Use this chart as a guide as you design activities, projects, and STEM challenges for the students you serve.

According to Hammond, teachers can use the features between individualism and collectivism to understand students' actions and attitudes. For example, a student from a collectivistic society—such as Pakistan, Ethiopia, or Guatemala—may act in ways that emphasize relationships, interdependence within a community, and cooperative learning. A student from an individualistic society—such as France, Belgium, or the United States—may behave in ways that emphasize individual achievement and independence.

In *Cultures and Organizations: Software of the Mind*, Geert Hofstede, Gert Jan Hofstede, and Michael Minkov (2010) discuss the following differences between collectivist and individualist societies in language, personality, and behavior.

- Collectivist societies:
    - Avoid using the word *I*
    - Score more introverted on personality tests
    - Encourage showing sadness and discourage showing happiness
    - Walk more slowly
    - Trust the social network as the primary source of information
    - Spend a smaller share of private and public income on health care
- Individualist societies:
    - Encourage using the word *I*
    - Score more extroverted on personality tests
    - Discourage showing sadness and encourage showing happiness
    - Walk faster
    - Trust media as the primary source of information
    - Spend a larger share of private and public income on health care

Collectivism and individualism refer to the extent to which individuals prioritize group goals over personal goals or vice versa, which can vary widely across different cultures. Understanding collectivism and individualism is crucial in teaching because these cultural dimensions significantly influence students' learning behaviors, their interactions with peers and teachers, and their overall educational expectations.

Hofstede and colleagues (2010) discuss the following key differences between collectivism and individualism as it pertains to school and the workplace.

- Collectivist societies:
  - Rely on the group to select students to speak up in class
  - See the purpose of education as learning how to *do*
  - Have lower occupational mobility
  - Feel that direct appraisal of subordinates spoils harmony
- Individualist societies:
  - Expect students to speak up individually in class
  - See the purpose of education as learning how to *learn*
  - Have higher occupational mobility
  - Feel that management training should teach the honest sharing of feelings

These dynamics can help you understand that students' actions are informed by their culture and avoid misunderstanding their actions as disrespectful or disinterested.

Understanding individualism or collectivism can greatly improve how educators design STEM lessons to create a learning environment where every student feels valued and motivated to participate. It will not only demonstrate respect for different cultures, but also allow integration of diverse perspectives into lessons that will prepare students to thrive in a globalized world. Educators can create more inclusive, supportive, and effective learning environments by integrating an understanding of collectivism and individualism in teaching practices. This cultural sensitivity respects and validates the diverse student backgrounds and enhances educational practices by appropriately addressing all students' needs and expectations.

Culture Informs Behavior

Culturally responsive teaching is *not* about race, it is about culture. *Education Week* reporters Madeline Will and Ileana Najarro (2022) give an example of culturally responsive teaching using White students to illustrate the focus is not on race. "In a predominately White school district, there are White students who, due to where they live or their family's socioeconomic status, are also underserved by their school district and could benefit from a culturally responsive approach to education." Why? According to Sharroky Hollie (2018), a national educator who provides professional development to thousands of educators in cultural responsiveness, this approach is more than simply incorporating a student's racial background into the classroom or thinking of students in a one-dimensional, stereotypical way. Culturally responsive teaching involves having teachers consider a student's gender, age, socioeconomic status, and whether they live in a rural community or the suburbs (Will & Najarro, 2022). It is about valuing the culture that all students bring to the classroom, regardless of race.

Figure 3.1 illustrates the *cultural iceberg*, the levels of culture that students bring with them into the classroom.

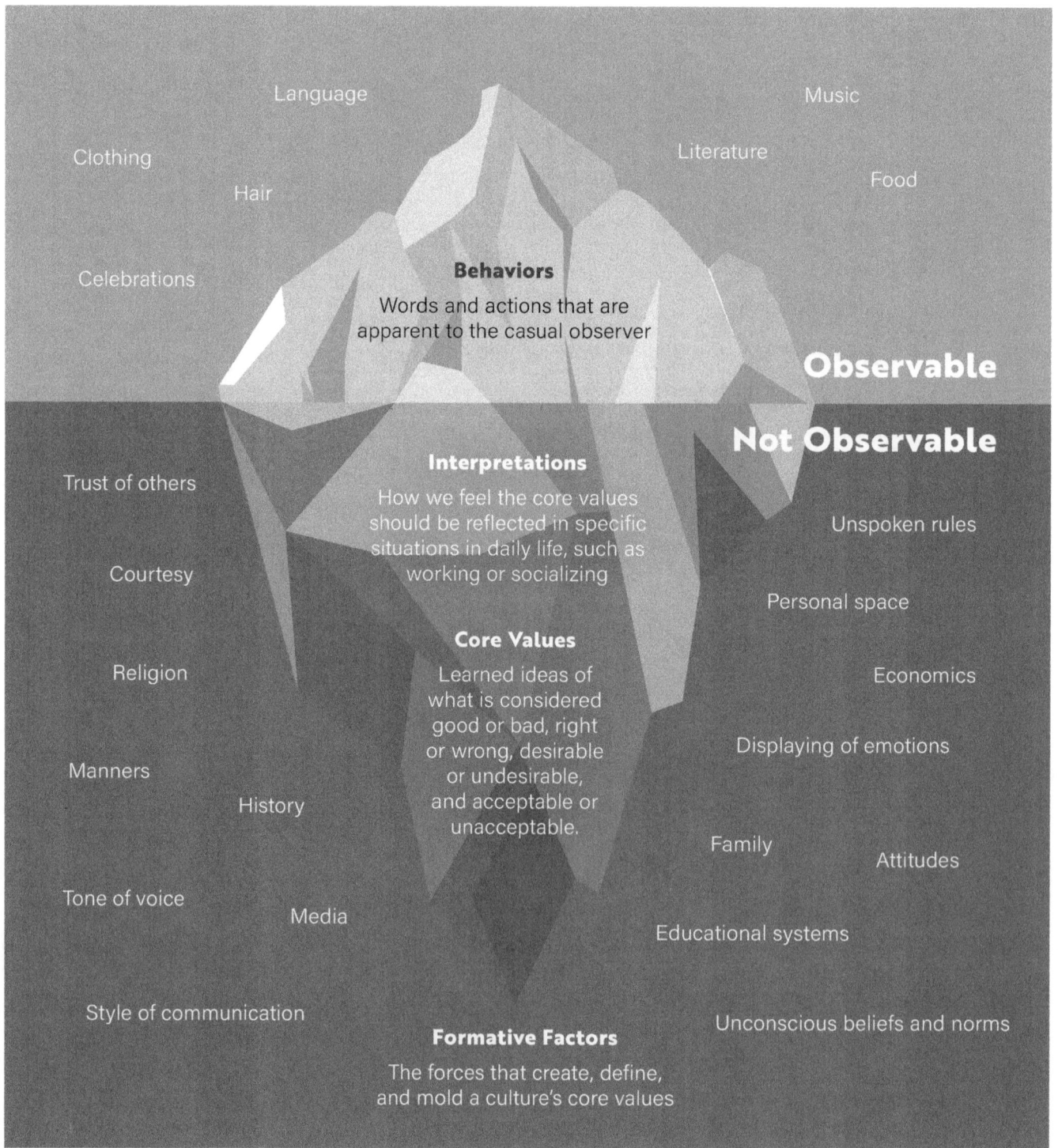

FIGURE 3.1: The cultural iceberg.

The iceberg has three layers: (1) behaviors, (2) interpretations, and (3) core values. Behaviors are observable: hair, clothing, speech styles, and so on. Interpretations and core values are deep cultures that you cannot see by looking at the student but define how they will interact with others. Although the examples from *Education Week* are students from the same race, their unobservable cultures can vary due to different socioeconomic classes. This area of unobservable cultures defines the students' beliefs, norms, how they interact with others, and where their trust lies, as examples.

Icebergs, like students, are greatly affected by their environment—they change and shift over time based on their surrounding environment. Zaretta Hammond (2015) explains that deep culture "is what grounds the individual and nourishes their mental health. It is the bedrock of self-concept, group identity, approaches to problem solving and decision making" (pp. 22–24). So, remember that culturally responsive teaching is not about race, it is about culture—all three of its layers!

Implicit Bias as a Barrier

Rucker (2019) proposes that to be culturally responsive, teachers need to take a deep look inside themselves (the personal dimension) and ask hard questions like the following (Erdner, 2020).

- Am I operating from a place of critical care within my classroom—a place that marries high expectations with empathy and compassion?
- Am I holding my students, regardless of socioeconomic status or background, to high standards?
- Does my past interaction with a particular race of people impact my ability to communicate with families or students?

These questions aim to make teachers aware of any *implicit bias* (also called blind spots)—a negative attitude, of which one is not consciously aware—influencing their relationships with students (American Psychological Association, n.d.). Use the QR code to learn more about blind spots.

Hammond's (2015) approach differs from Rucker's (2019), arguing that addressing implicit bias is not the first step in culturally responsive teaching—rather, the first step is developing relationships with your students.

Blind Spots
www.pwc.com
/us/en/about-us
/blind-spots.html

I believe that the two must occur in tandem if we, as teachers, are to culturally diversify STEM classrooms because it's difficult to cultivate relationships if the teacher holds biases that compromise students' trust. Therefore, it is imperative for educators to identify the places in their instructional planning or in their interactions with students where bias prevents them from supporting students to achieve at optimal levels.

Consider the following questions to determine how bias might affect your instructional planning.

- What is the basis for determining who gets rigorous work and who doesn't?
- When students are answering questions, do you avoid calling on certain students because you expect lower-quality responses?
- Are your expectations of students creating dependent learners or independent learners?
- Which students are your dependent learners and why?
- Which students are your independent learners and why?

Consider the following questions to determine how bias might affect your interactions with students.

- What does your body language communicate to students?
- Do you avoid or minimize interactions with specific students?
- When you grade student work, are you harsher on some students? Are you shocked when some students do well?
- Do you ever feel a need to double check just to make sure you graded a student's work properly?

Take a moment to reflect on the preceding questions and be honest with your responses. Although your answers may be difficult to accept, understand that your life experiences may have led you to develop stereotypes and biases. This bias, as unintentional as it may be, can impact how you relate to students and their families (Rucker, 2019). This reflection will also serve as a framework as you read each of the student scenarios presented in this book and answer the corresponding questions.

Your reflections may serve as your first step in understanding your actions or inability to create change in your classroom and instructional practices for multilingual learners. If you commit to becoming a culturally responsive teacher, students will feel a sense of safety, respect, and belonging in your classroom (Rucker, 2019).

Meet Maria

Meet Maria Christina Torres
https://youtu.be/UlVArgLfg
E0?si=7TCZ_-2mlONZzdV5

It's time to meet your first multilingual student. Her name is Maria Christina Torres. She arrived in the United States during the summer and will be in your class starting the first day of school in September. Scan the QR code to view a message from Maria.

You will receive background information on Maria in increments, as it is common for teachers to receive information about multilingual learners in bits and pieces over time. Very rarely does everything come at once. You will also encounter reflection questions along the way. Use these moments of reflection to guide your journey into culturally responsive teaching.

During Maria's first month at school, you experience several of the challenges common to multilingual learners (chapter 2, page 25). Maria struggles to communicate due to the language barrier, which impacts your ability to give her instructions, explain concepts, and assess her understanding. Cultural differences are also apparent as you struggle to understand how Maria learns best and interacts with her peers. It's been challenging to determine an adequate pace for her to have enough time to process information while ensuring that the other students remain engaged and challenged.

Figure 3.2 (page 51) contains Maria's bio.

After this initial meeting with Maria, you reflect on the experience so far and consider the following questions.

- What views or thoughts come to mind about Maria?
- What assumptions have you made about her family and their views on education?
- How do you feel about having Maria in your class (for example, excited, stressed, or worried)?
- What are your concerns (such as her ability to keep pace with English-speaking students, the potential language barrier, and so on)?
- How will you introduce her to your class?
- Do you foresee any cultural conflicts between Maria and her peers?
- What are your academic expectations for Maria? What are your expectations based on?

Over the next few weeks, you and Maria continue to experience challenges; you both struggle to be patient with the process of her acclimation to life in a U.S. public school. For Maria, the honeymoon stage of culture shock is coming to an end, and she is struggling to keep up with the work and to understand the lessons. The school and her class are larger than she's used to, and she is beginning to miss back home. You've been seeking resources, training, and support from more experienced colleagues and organizations dedicated to teaching multilingual learners, but you still struggle with keeping the English speakers in class engaged while tending to Maria's needs. You also find it challenging to meet her needs without having access to her educational information.

One month after Maria arrived in your class, you finally receive additional information about her language proficiency and academic achievement according to her records, shown in figure 3.2 (page 51). This information will help you understand what Maria's general capabilities are in the areas of reading, writing, speaking, and listening. Refer to chapter 2 (page 25) as needed to review the language acquisition stages to better understand Maria's language proficiency. Now you will be able to better support Maria academically to minimize misunderstandings and frustrations between both parties.

Maria Christina Torres

Maria Christina Torres was born in Los Bordos, an impoverished town in Ecatepec de Morelos, deemed one of the most dangerous cities in Mexico where at least six hundred women have been murdered since 2012. It is located just half an hour from the capital. Los Bordos has very little running water or electricity, few schools, and fewer hospitals. While most young children attend primary school, only half of the students who begin primary education finish primary school, and seven out of ten teens do not master reading comprehension or mathematics. Of the 62 percent of students who reach secondary school, only 45 percent graduate, in contrast to about 75 percent of U.S. students who graduate high school on time with a regular diploma. Maria was in the seventh grade when she emigrated from Mexico to the United States with her mother and younger sister. Maria's parents were divorced, and she had little contact with her father. The Torres family moved in with Maria's aunt, who had come to the United States from Mexico ten years earlier and lives in a two-bedroom apartment. As a little girl, Maria's nickname had been Luna, which is the Spanish word for moon. She had dreamed of one day traveling into space and maybe even visiting the moon.

Maria had been an excellent student in Mexico and had read everything she could locate in the local library about astronaut Ellen Ochoa, the first Hispanic woman to go to space. Ochoa had served on a nine-day mission aboard the Space Shuttle Discovery, where she and a team of astronauts studied the Earth's ozone layer. Ochoa returned to space three more times, spending nearly one-thousand hours in orbit. Today, she holds the NASA Distinguished Service Medal and serves as the Director of the Johnson Space Center in Houston, Texas.

Maria speaks limited English, and Spanish is the primary language spoken in her home. The school that Maria attended in Mexico had fewer than three hundred students, so despite the language difference, she was comfortable in a small school.

The public education system remains one of the most arduous challenges that Mexico faces. Mexico is the largest Spanish-speaking country in the world. According to the Programme for International Student Assessment (PISA), a worldwide study by the Organisation for Economic Co-operation and Development (OECD), fifteen-year-old Mexican students' scholastic performance on mathematics, science, and reading ranked last among students in the thirty-six OECD member countries in 2018 (MND Staff, 2019).

Name	Country of Birth	Primary Language	Level of Language Proficiency	Age	Grade
Maria Christina Torres	Mexico	Spanish	Level 1–Level 2: Early Production or Emerging	Thirteen	Sixth
WIDA Results					
Listening: 2.0		Speaking: 2.0		Reading: 1.0	Writing: 1.0
Student Capabilities By Level					
Listening (proficiency level 2): Students can understand oral language related to specific familiar topics in school and can participate in class discussions. They can identify main topics in discussions, categorize or sequence information presented orally using pictures or objects, follow short oral directions with the help of pictures, and sort facts and opinions stated orally.					
Speaking (proficiency level 2): Students can communicate ideas using words and phrases related to everyday routines or situations. They can retell stories or content-related events; state procedural steps; give reasons why or how something works using diagrams, charts, or images; and state opinions based on experiences.					
Reading (proficiency level 1): Students can understand written texts that include visuals and may contain a few words or phrases in English. This includes interpreting information from graphs, charts, and other visual information; identifying steps in processes presented in graphs or short texts with illustrations; identifying words and phrases that express opinions and claims; and comprehending short text with illustrations and simple and familiar language.					
Writing (proficiency level 1): Students can communicate in written English using language related to familiar topics in school. This includes describing ideas or concepts using texts and illustrations, labeling steps in processes presented in graphs or short texts, stating opinions or preferences through text and illustrations, and sharing personal experiences through drawings and words.					

Source: Student bio adapted from J. Kellmayer, personal communication, November 2021.

FIGURE 3.2: Maria Christina Torres's bio and academic profile.

After another month of working with Maria and receiving her academic background information, you stop and reflect on the process. You consider the following questions and notice how your responses at this point compare to last month's responses.

- Now that you have more background information about Maria's language and academics, would you change any of your previous responses? Which one and why?
- Do you notice any bias in your responses now that you have additional information on Maria?
- What assumptions are you making about Maria based on the language and academic information provided?
- What concerns do you have regarding your self-efficacy in teaching Maria?
- What support would help you be successful in teaching Maria and addressing her needs as an English learner?

Culturally Responsive Teaching in Practice

Maria has now been in your classroom for a little over two months. You have received her WIDA scores and have been able to make some modifications to assignments. But it is still challenging to balance her needs with the pacing you need to maintain to complete the units as written in the curriculum. You are falling behind in the pacing guide and you're feeling overwhelmed, even though you've reached out to the English language instruction and bilingual teachers for assistance and spoken with your building administrator.

You begin noticing a difference in Maria, and you are concerned. Maria is becoming frustrated as she struggles to master academic vocabulary, understand the language, and be accepted into a STEM class. Maria's initial interest and ability in science starts to wane; she becomes withdrawn, frustrated, and depressed. She resists participating in class activities.

You recognize that Maria is experiencing the frustration stage of the four stages of culture shock (chapter 2, page 28), often characterized by irritability, anxiety, frustration, and hostility. Remembering that this stems from a students' inability to understand language, signs, and cultural norms, you don't take it personally. You know that after this rejection stage passes, Maria can begin to feel more relaxed and develop a more balanced view of her life in the United States *if* all parties (teachers, counselors, advisers, and family members) recognize and address this stage's challenges. You hold realistic expectations for this process, remembering that students can reenter culture shock, especially if they return to their country of origin for a visit.

Given what you've learned about multilingual learners' needs and challenges in general and Maria's situation in particular, how do you support her at this time? And how does what you've learned in this chapter about culturally responsive teaching inform your approach? The following section provides possible practices for answering these questions.

Understand Language Development and Proficiency

It's important to understand what Maria's level of language acquisition (chapter 2, page 30) represent about her abilities to interact in the classroom. Consider the following points.

- Since Maria is in the early production stage of language acquisition, you know that she has limited comprehension without support and will most likely answer in present tense with one- or two-word sentences.

- You know that Maria can respond to yes-or-no questions; either-or questions; and who, what, and how many questions.

- Based on her WIDA proficiency results in speaking and listening, you can expect Maria to communicate ideas using words and phrases related to everyday situations and understand messages or directions involving language related to familiar experiences.

- Visuals are necessary in all aspects of teaching Maria.

Knowing this information provides you with an idea of how to communicate with this student and create grade-level assignments and assessments for her. It also provides valuable information about how to expect Maria to respond to you. If possible, have a translator with you when you speak with Maria, discuss your observations with her parents, and express your concerns based on her actions. If this option isn't available in your district (it's not available in most districts), consider using a free translation application to translate documents or convert text to speech. The following are great options to get you started.

- Google Translate
- iTranslate
- iTranslate Voice
- SpanishDictionary.com Spanish Translator
- Microsoft Translator
- SayHi Translate
- Apple Translate
- Speak and Translate
- DeepL

Create a Culturally Responsive Environment

You spend time researching Mexican culture to learn more about Maria. On the individualism-collectivism continuum, Mexico has a score of thirty. You are, therefore, not surprised to learn the following (Johnson & Rodriguez, 2005).

- Warm relationships between teachers and students are common, which in Mexico is an extension of the parent-child relationship, especially in preschool and primary grades (de Souza & Lee, 2017). This dynamic is based on mutual respect between teachers, students, and parents. Respect for teachers is a cultural value in Mexico.

- The treatment of adults and superiors, including parents and teachers, is described as a tradition of respect and submission. The most common misunderstanding prompted by this cultural difference is when a student lowers their eyes in the presence of an adult or an authority figure. This mandatory behavior taught in Mexico as a sign of respect may be interpreted as an admission of guilt in the United States.

- Behaviors shaped by cultural differences and fear of lawsuits may lead Mexican students to believe that teachers in the United States do not like them.

- For most Latinos, including Mexicans, personal space—the culturally determined distance maintained between people—is much smaller than for White Americans (Evason, 2018). Mexican children and adults will misinterpret your maneuvering to keep a "comfortable" space between you and them.

- Because Secundaria (the equivalent of seventh, eighth, and ninth grades) has only been required in Mexico since 1993, and school attendance is only required until fourteen years of age, Maria's parents may not be aware of the compulsory attendance laws in the United States nor understand the expectation that everyone should complete high school.

In addition to completing personal research, you turn to school-based resources (a guidance counselor or a bilingual teacher on your campus) for information about Maria's background. You ask for details about the city she moved from as well as her home-life situation, siblings, and other background information to get as much context as possible. You feel that having a well-rounded picture of Maria's life will help you successfully meet her academic needs.

By understanding these aspects of Mexican culture and Maria's life, you can be better equipped to interact with Maria and her family in positive and supportive ways. It may even help you understand her parents' actions. Since she is thirteen, you may want to be sure that Maria and her parents understand she needs to stay in school to comply with U.S. attendance laws. This should also be a catalyst for you to do everything you can to help Maria through this phase of culture shock and future stages that may develop over time.

Keep Maria Engaged

You create an all about me questionnaire, like the one in figure 3.3, and administer it to all students. You've already confirmed that Maria can read Spanish, so you translate her questionnaire into Spanish, and provide a copy to her parents. The questionnaire will allow you to learn what Maria enjoys doing, what she likes to read, what she likes about STEM, what her favorite science topic is, and what she wants to learn more about in science class.

All About Me	
Todo Sobre Mí	
1. What is your full name?	
1. ¿Cuál es tu nombre completo?	
2. Who is your hero?	
2. ¿Quién es tu héroe?	
3. If you could live anywhere else, where would it be?	
3. Si pudieras vivir en otro lugar, ¿dónde sería?	
4. How many siblings do you have?	
4. ¿Cuántos hermanos tienes?	
5. How many places have you visited, and which one was your favorite?	
5. ¿Cuántos lugares has visitado y cuál fue tu favorito?	
6. How many languages do you speak?	
6. ¿Cuántos idiomas hablas?	
7. What makes you really angry?	
7. ¿Qué te hace realmente enojar?	
8. What makes you really happy?	
8. ¿Qué te hace realmente feliz?	

9. What motivates you to work hard? 9. ¿Qué te motiva a trabajar duro?	
10. Do you play sports? Which one is your favorite? 10. ¿Practicas algún deporte? ¿Cuál es tu favorito?	
11. Where were you born? 11. ¿Dónde naciste?	
12. What is your favorite book? 12. ¿Cuál es tu libro favorito?	
13. Who is your best friend? 13. ¿Quién es tu mejor amigo?	
14. Do you have any pets? If yes, what is their name and what type of pet is it? 14. ¿Tienes mascotas? Si sí, ¿cómo se llaman y qué tipo de mascota son?	
15. What is your favorite food? 15. ¿Cuál es tu comida favorita?	
16. What makes you unique? 16. ¿Qué te hace único?	
17. What is your favorite color? 17. ¿Cuál es tu color favorito?	
18. If you could buy anything, what would it be and why? 18. Si pudieras comprar cualquier cosa, ¿qué sería y por qué?	
19. What do you like to do in your free time? 19. ¿Qué te gusta hacer en tu tiempo libre?	
20. What food do you like least? 20. ¿Qué comida te gusta menos?	

FIGURE 3.3: All about me questionnaire.

*Visit **go.SolutionTree.com/EL** for a free reproducible version of this figure.*

You discover the following information about Maria from talking to her parents, teachers, and guidance counselors and from reading her answers to the all about me questionnaire.

- Maria Christina Torres was in the sixth grade when she emigrated from Mexico to the United States with her mother and younger sister.
- Maria's parents are divorced, and she has little contact with her father.
- Maria's family moved in with Maria's aunt, who came to the United States from Mexico ten years ago and lives in a two-bedroom apartment in rural western Texas.

- As a little girl, Maria's nickname was Luna, which is the Spanish word for *moon*. She dreamed of one day traveling into space and maybe even visiting the moon. She would often say, "Luna en la luna" (Luna on the moon).
- Maria was an excellent student in Mexico and read everything she could locate in the local library about astronaut Ellen Ochoa.
- Maria speaks limited English (defined as early production or speech emergence). Spanish is the primary language spoken in her home.
- The school that Maria attended in Mexico has fewer than five hundred students.
- Maria diligently works at improving her English but struggles with the academic language needed to be successful in class.
- Maria is feeling homesick. She misses her old school, where she was popular and had many friends.
- When Maria lived in Mexico, she was constantly surrounded by a large, extended family.
- Maria was very upset that she was not accepted to the STEM classes due to her lack of English proficiency.

You know Maria better now that you understand her living conditions; changes in climate, food, and so on; combined with the language barrier, nostalgia for relatives and friends, and differences in the schools that make life difficult for a binational youngster.

After considering all you've learned about Maria, answer the following questions. Also notice how your responses at this point compare to your previous responses.

- How will you help Maria become successful in your class with this new information?
- How would knowing language acquisition levels help you plan instruction for Maria?
- How does understanding language proficiencies enable you to communicate better with Maria?
- What can you do to make Maria feel like a member of the class?
- What strategies would you use with your new knowledge of multilingual learners?
- Did your perception of Maria change as you learned more about her? In what way?
- Had you known this information about Maria earlier, do you think your responses to the previous reflection questions in this chapter would have been different? Why or why not?

Having Maria and the class complete an all about me questionnaire is beneficial because it fosters a sense of community and understanding within the classroom between Maria and her new peers. This activity allows students to share their backgrounds, interests, and unique aspects of their identities, which helps teachers and classmates learn more about each other in a personal and meaningful way. It also provides teachers with valuable insights into their students' lives, which can be used to tailor lessons and interactions to better meet individual needs and preferences. Moreover, for the students, especially Maria, this exercise can boost self-esteem and promote self-expression, encouraging students to feel more connected and engaged in the classroom environment. Overall, an all about me questionnaire is a simple yet powerful tool to enhance inclusivity, personalize learning, and build stronger relationships among students and teachers.

Implement Strategies and Recommendations

The following strategies may help you assist Maria and her family in becoming acculturated to their new environment and country.

- Teach the students how to pronounce a few words of greeting in Spanish.
- Collect resources about Maria's home culture and language and display them around the classroom.
- Highlight STEM professionals and famous figures from diverse backgrounds.
- Include multicultural literature or books in Maria's native language. This also helps the English-speaking students view diverse cultures as important contributors to STEM (Rodriguez & Bell, 2018).
- Make Maria's other teachers aware of the cultural differences that you learned, including the information you obtained from the individualism-collectivism continuum.
- Speak with the principal about having an orientation if Maria and her family were not provided one and include the following.
  - A tour of the school that helps Maria locate classrooms, bathrooms, the cafeteria, and the bus stop, and introduces her to key staff, including those who speak Spanish.
  - Finding her locker and learning how to use the lock.
  - Prepare an information packet for Maria's teachers, counselor, and administrators. This might include a description of Maria's family, hometown in Mexico, academic achievement, interests, and so on.
- Try your best to make her feel comfortable in the class by assigning her peer buddies, preferably those who are Spanish-speaking and Mexican.
- Provide Maria with a Spanish-English dictionary or an electronic multilingual dictionary, such as LingvoSoft Online.
- Familiarize yourself with the stages of culture shock (chapter 2, page 28) and inform your team so that they are also aware of the stages, how to identify it, and how to address it.

- Assign an autobiography project for students to complete but have them include their science interest. This will provide an opportunity for the students to learn about Maria and for Maria to learn about her classmates. This can also help her share her experiences and life in Mexico. Since Maria's level of language acquisition is at early production, speaking in front of the class may be difficult, so allow her to use a picture template to visually represent her thoughts, or a computer program, such as Flip, to make a movie or slideshow. This provides Maria with the opportunity to write, check, and practice what she wants to share with her classmates.

- Given Maria's levels of language proficiency and language acquisition, create lessons that will allow her to experience success in the STEM classroom. Revisit this chapter to see how you will use that information to teach Maria so she can be successful in your class.

Helping students and their families become acculturated is vital for fostering an educational environment that supports all aspects of a student's development—academic, social, and emotional. It is a critical step toward building inclusive school communities that respect and celebrate diversity while promoting equality and understanding.

The STEM Challenge

You've now had the opportunity to learn about Maria and reflect on any implicit biases, assumptions based on limited information, and ideas about multilingual students that may have affected your academic expectations of students whose first language is not English. In this section, I introduce a planning process you can use to design a STEM lesson. This process provides strategies specific to multilingual learners and considers a student's prior knowledge, funds of knowledge, and language proficiency. This approach will increase Maria's engagement and provide opportunities for her to make appropriate classroom connections using examples, comparisons, and strategies using a STEM challenge as the mechanism.

What is the difference between a STEM challenge and a traditional science lab? A STEM challenge is not a quick lab that is to be completed in one forty-five-minute class like a regular science lab. Both STEM challenges and traditional science labs offer hands-on learning experiences for students, but they differ in purpose, structure, and the skills they aim to develop. Table 3.4 illustrates the differences.

As you can see, both STEM challenges and traditional science labs have their merits. While regular science labs are essential for building foundational scientific knowledge and understanding the scientific method, STEM challenges are crucial for developing holistic STEM skills and preparing students for real-world applications.

Keep the following in mind when selecting a STEM challenge.

- The STEM challenge should be interactive and promote discourse.
- The STEM challenge should involve student action to demonstrate or provide insight into what they know and what they've learned about the topic.
- The STEM challenge should allow all students to connect with the material while aligning with the standard(s) you are teaching.

TABLE 3.4: STEM Challenge Versus Traditional Science Lab

Purpose	
STEM Challenge	Regular Science Lab
These are typically designed to solve real-world problems or challenges using the principles of science, technology, engineering, and mathematics. The focus is often on creativity, innovation, and applicability.	Traditional labs often aim to demonstrate a particular scientific principle, experiment, or procedure. The primary goal is to reinforce theoretical concepts discussed in lectures or textbooks.
Structure	
STEM Challenge	Regular Science Lab
These are more open-ended, allowing students to take different approaches to solve the given problem. They can involve a multidisciplinary approach where students draw on knowledge and skills from various STEM areas.	Labs often follow a structured procedure or set of instructions. Students are typically required to follow these steps to achieve the desired outcome or observation.
Skills Developed	
STEM Challenge	Regular Science Lab
These challenges tend to foster a broader range of skills, including solving problems, critical thinking, teamwork, design thinking, resilience, and adaptability.	Traditional labs primarily focus on skills related to the scientific method, such as observation, hypothesis formation, experimentation, and analysis.
Outcome	
STEM Challenge	Regular Science Lab
These STEM challenges can have multiple acceptable solutions or outcomes given their open-ended nature, and students might come up with a variety of innovative solutions.	Labs usually have a specific expected outcome or result based on the scientific principle being demonstrated or studied.
Resources	
STEM Challenge	Regular Science Lab
These challenges often require a variety of materials, potentially from different STEM disciplines, to allow for diverse solution strategies. This might include anything from building materials to coding platforms.	Traditional labs generally use specific equipment and materials designed to demonstrate a particular scientific principle or experiment.
Duration	
STEM Challenge	Regular Science Lab
These challenges can be short term or extend over a longer period, such as several days or even weeks, as students iterate and refine their solutions.	They are typically completed in one class period or lab session.

The next section discusses guidance for planning the challenge. The STEM challenge that will be used involves something common that most students either have seen or been in—an airplane. But it is not about explicitly telling students why airplanes fly; instead, it's about having them determine how to get a paper airplane to glide the farthest while carrying a load. This will allow the students to use everything they know and think they know about why airplanes fly by doing something many students have done no matter where they live—making paper airplanes. Teachers often make the mistake of answering students' questions by telling them what people have already figured out. Instead, STEM challenges allow students to start with their preexisting experiences, then the challenge walks them through the concepts. Their experiences will lead them to answer the challenge question—in this case, how to get the airplane to glide the furthest with the heaviest load. Keep in mind that the brain wants to make sense of the general information before the details. This STEM challenge provides students with opportunities to solve problems, think critically, and use old knowledge to solve a new problem.

Plan the STEM Challenge

When planning a STEM challenge based on culturally responsive teaching, it may be helpful to begin with the five essential components of culturally responsive teaching (Will & Najarro, 2022). According to these components, teachers should have the following:

- **A strong knowledge base about cultural diversity:** Teachers should understand different racial and ethnic groups' cultural values, traditions, and contributions to society, and incorporate that knowledge into their instruction.
- **Culturally relevant curricula:** Teachers should include multiple perspectives in their instruction and make sure the images displayed in classrooms—such as on bulletin boards—represent a wide range of diversity. Teachers should also contextualize issues within race, class, ethnicity, and gender.
- **High expectations for all students:** Teachers should help students achieve academic success while still validating their cultural identities.
- **An appreciation for different communication styles:** Teachers should understand different communication styles and modify classroom interactions accordingly. For example, many communities of color have an active, participatory style of communication. A teacher who doesn't understand this cultural context might think a student is being rude and tell the student to be quiet. The student may then shut down.
- **The use of multicultural instructional examples:** Teachers should connect students' prior knowledge and cultural experiences with new knowledge.
(Will & Najarro, 2022)

Overall, culturally responsive teaching methodology involves incorporating students' cultural references in all aspects of learning, which enhances student engagement and makes the educational experience more relevant and effective. Teachers practicing this approach are committed to learning about the cultural backgrounds of their students and fostering a classroom environment that respects and celebrates these differences. They also adapt their teaching strategies to accommodate varied learning styles and backgrounds, ensuring that all students have equitable access to education. Culturally responsive teaching not only supports academic achievement, but also helps students develop a positive self-concept by affirming their cultural identities in school settings. "As Emily Style, the former founding co-director of the National SEED Project (Seeking Educational Equity and Diversity), once wrote, 'Half the curriculum walks in the door with the students'" (Will & Najarro, 2022).

The next tool to draw from during the planning process is the culture and language framework, a tool adapted from Zaretta Hammond's work. Hammond (2015) writes that teachers must first understand the basic concepts of culturally responsive pedagogy before they can learn the instructional moves associated with them. Hammond (2013) created a Ready for Rigor framework designed to help teachers with instructional moves by organizing key areas of teacher capacity building. This framework helps teachers move students from dependent learners (the teacher carries most of the task's cognitive load) to independent learners (the learner has control and ownership of their learning and can make informed choices, set goals, and make decisions about how to fulfill their learning needs; Main, 2022). The framework consists of four areas of culturally responsive teaching: (1) awareness, (2) learning partnerships, (3) information processing, and (4) community building.

Figure 3.4 (page 62) uses the Ready for Rigor framework as a starting point, incorporating language learning strategies as they apply to multilingual learners trying to learn content and language simultaneously while understanding culture's importance in teaching culturally diverse students.

The culture and language framework make the connection between language and culture that teachers must consider when planning lessons. Use the framework as a reminder of key elements of culturally responsive teaching you should be aware of when planning, such as implicit bias, language development, and proficiency levels. It keeps you in tune with research on the importance of collaboration in academic language development and how your classroom environment sets the tone in student engagement, success, and failures.

Awareness	Learning Partnerships	Information Processing	Community of Learners and Learning Environments
Teachers should be aware of: • The levels of culture as seen in the cultural iceberg (behaviors, interpretations, and core values; see figure 3.4, page 46) • The brain's involvement in learning and actions that can trigger the fight, flight, or freeze reactions and what that looks like in the classroom • The identification of cultures as independent or collective and how this drives instruction activities • How the teacher's culture resembles or differs from that of students and how that difference can create a deficit lens or a false perception of students • Language development; as multilingual learners begin, their language will be less sophisticated but that does not mean that they cannot contribute to the conversation	Teachers should understand: • The importance of student and teacher relationships • The impact of microaggressions and implicit bias on a students' social-emotional well-being and academic achievements • The benefit of student-to-student relationships and how learning is a social endeavor that requires a high level of interaction and making connections • The importance of student collaboration in developing language skills that can strengthen academic language • That the roles of teacher and student are interrelated and that both parties should take active part in the learning process	Teachers should recognize the importance of: • Connections between content and students' everyday lives to make learning relevant • Using multiple modalities in teaching and allowing students and yourself to use gestures, words, and objects as resources to explain, teach, or understand the content • Understanding the levels of language acquisition in planning instruction for content, process, and product • Supporting student participation in classroom discussions by rephrasing, asking for clarification, probing student thinking, and building on what students say • Understanding the brain's working memory capacity	Teachers should create: • An environment where student identities are respected and embraced • A safe environment by making English learners feel accepted, respected, and included in the classroom • Assessments where students have multiple opportunities to demonstrate their understanding in a variety of ways • A classroom that includes multicultural literature in your English learners' native languages • An environment that engages students in productive discourse and interactions with others

FIGURE 3.4: The culture and language framework.

IDENTIFY THE OBJECTIVE AND STANDARD

It's time to begin planning your STEM challenge. We begin by identifying the objective and the standard before determining the goals, action steps, vocabulary, and hands-on activity for each phase of the STEM challenge.

Figure 3.5 (page 64) contains a sample STEM challenge.

The standards for this STEM challenge may also be adapted for high school students, all the way to grade 12. Consider the Next Generation Science Standards for grades 9–12 (Achieve, Inc., 2017):

> The high school performance expectations in Physical Science build on the middle school ideas and skills and allow high school students to explain more in-depth phenomena central not only to the physical sciences, but to life and earth and space sciences as well. These performance expectations blend the core ideas with scientific and engineering practices and crosscutting concepts to support students in developing useable knowledge to explain ideas across the science disciplines. In the physical science performance expectations at the high school level, there is a focus on several scientific practices. These include developing and using models, planning, and conducting investigations, analyzing and interpreting data, using mathematical and computational thinking, and constructing explanations; and to use these practices to demonstrate understanding of the core ideas. Students are also expected to demonstrate understanding of several engineering practices, including design and evaluation. (p. 1)

The following are the steps you might take to plan a STEM challenge. This STEM challenge serves as an example of the steps.

- Know the vocabulary and goals for the entire STEM challenge.
  - *Vocabulary*—This challenge includes terms like weight, thrust, drag, lift, low air pressure, high air pressure, gravity, cargo, control, variables.
  - *Goal*—Students should be able to design a paper airplane that will glide the farthest carrying a cargo load (pennies), and to make sense of the vocabulary through hands-on activities.
- Determine which science domain(s) to focus on and select those standards that are in alignment with the STEM challenge. For example, for this challenge involving flight, I looked at the physical science standards and reviewed the explanation of the standard for each grade band that was below and above my targeted grade level. I then reviewed the mathematics standards listed in the Next Generation Science Standards document to know what mathematics skills each student needs and should have been taught (for example, measurement, conversions, and so on).
- Plan and chunk the hands-on activities accordingly.

Stem Challenge: Make a paper airplane that can carry a cargo load and glide more than ten feet.

Science: Bernoulli's Principle, the four forces of flight, and how airplanes fly

Standards:

Disciplinary Core Ideas (DCI):		
Next Generation Science Standard	Grade Level	Explanation of Standard
MS-PS2-2 Motion and Stability: Forces and Interactions	Middle school: grades 6–8	Plan an investigation to provide evidence that the change in an object's motion depends on the sum of the forces on the object and the mass of the object.
MS-ETS1-2 Engineering, Technology and the Application of Science	Middle school: grades 6–8	Evaluate competing design solutions using a systematic process to determine how well they meet the criteria and constraints of the problem.
MS-ETS1-4 Engineering, Technology and the Application of Science	Middle school: grades 6–8	Develop a model to generate data for iterative testing and modification of a proposed object, tool, or process such that an optimal design can be achieved.
5-PS-1 Motion and Stability: Forces and Interactions	Elementary school: grade 5	Plan and conduct an investigation to provide evidence of the effects of balanced and unbalanced forces on an object's motion.
5-PS1.A Structure and Properties of Matter	Elementary school: grade 5	Matter of any type can be subdivided into particles that are too small to see, but even then the matter still exists and can be detected by other means.
5-PS2.B Types of Interactions	Elementary school: grade 5	The gravitational force of Earth acting on an object near Earth's surface pulls that object toward the planet's center.
3-5-ETS1-1 Engineering, Technology and the Application of Science	Elementary school: grades 3–5	Define a simple design problem reflecting a need or a want that includes specified criteria for success and constraints on materials, time, or cost.
3-5-ETS1-2 Engineering, Technology and the Application of Science	Elementary school: grades 3–5	Generate and compare multiple possible solutions based on how well each is likely to meet the criteria and constraints of the problem.
3-5-ETS1-3 Engineering, Technology and the Application of Science	Elementary school: grades 3–5	Plan and carry out fair tests in which variables are controlled and failure points are considered to identify aspects of a model or prototype that can be improved.
Crosscutting concepts: Cause and effect, systems, and systems models		
Science and Engineering Practices: • Practice 2: Developing and using models • Practice 3: Planning and carrying out investigations • Practice 5: Using mathematics and computational thinking (graphing)		

Source: Adapted from NGSS Lead States, 2013.

FIGURE 3.5: Sample STEM challenge.

Figure 3.6 shows the hands-on activities I chose for this STEM challenge. They have two purposes: (1) build and assess students' background knowledge and (2) build vocabulary through sensemaking. To view detailed instructions for each activity, visit this book's webpage at **go.SolutionTree.com/EL**.

Goal	Vocabulary	Activity
Determine how weight impacts flight	• Gravity • Weight • Cargo	Roto-copter
Experience Bernoulli's Principle	• Lift • Low pressure • High pressure	Paper strip
Learn to work like a scientist, keeping track of results and how to experiment and control variables	• Controls • Variables	Spinning blimps
Understand the rudder and how it controls the position of the nose of the aircraft *not* to turn the plane	• Rudder	Fabulous foam flier

FIGURE 3.6: Sample goals worksheet for planning a STEM challenge.

*Visit **go.SolutionTree.com/EL** for a free reproducible version of this figure.*

Through this approach, students learn about the fundamental properties of matter and see how these properties play a crucial role in everyday phenomena like flying. This reinforces the scientific concepts and enhances their understanding of how science applies to real-world scenarios. When considering standards for a STEM challenge, be sure to look at the standards above and below the grade level you teach. This allows you to make connections with the students' prerequisite knowledge and provide just enough challenge to get students thinking.

SET GOALS AND ACTION STEPS

Break down the challenge into phases so that the information is not overwhelming for students; this also allows you to plan the challenge in stages. Review the entire challenge, then determine the goals, action steps, vocabulary, and hands-on activity for *each* phase of the challenge as a part of the engineering design process.

The *engineering design process* is a series of steps that engineers follow to find a solution to a problem. The steps for the engineering design process depend on the project, but it provides opportunities for students to learn from their failures and to make improvements to their design. The engineering design process is an iterative process, where students can repeat steps to allow for improvements. This process consists of seven key steps:

1. **Identify the problem:** Clearly define the issue or need that requires a solution.
2. **Research or gather information:** Collect relevant data, research similar existing solutions, and determine the limitations for the solution.

3. **Brainstorm ideas:** Generate a wide range of potential solutions without immediate judgment, fostering creativity.
4. **Create a plan:** Develop a detailed plan or blueprint for the most promising idea, considering necessary resources and limitations.
5. **Create a prototype:** Construct a preliminary version of the solution to test its feasibility.
6. **Test prototype:** Assess the prototype's performance against the initial criteria, documenting its strengths and weaknesses.
7. **Analyze results and modify:** Review the test data, identify areas for improvement, and refine the solution through modifications or by repeating previous steps.

Figure 3.7 illustrates the steps. Notice the double arrows between "gather research and brainstorm," "create a plan and create a prototype," and "test prototype and improve or repeat." This is because these are phases in the engineering design process where students will either make iterations to their design or go back and research more information to tackle the problem and create a successful prototype that addresses the identified problem.

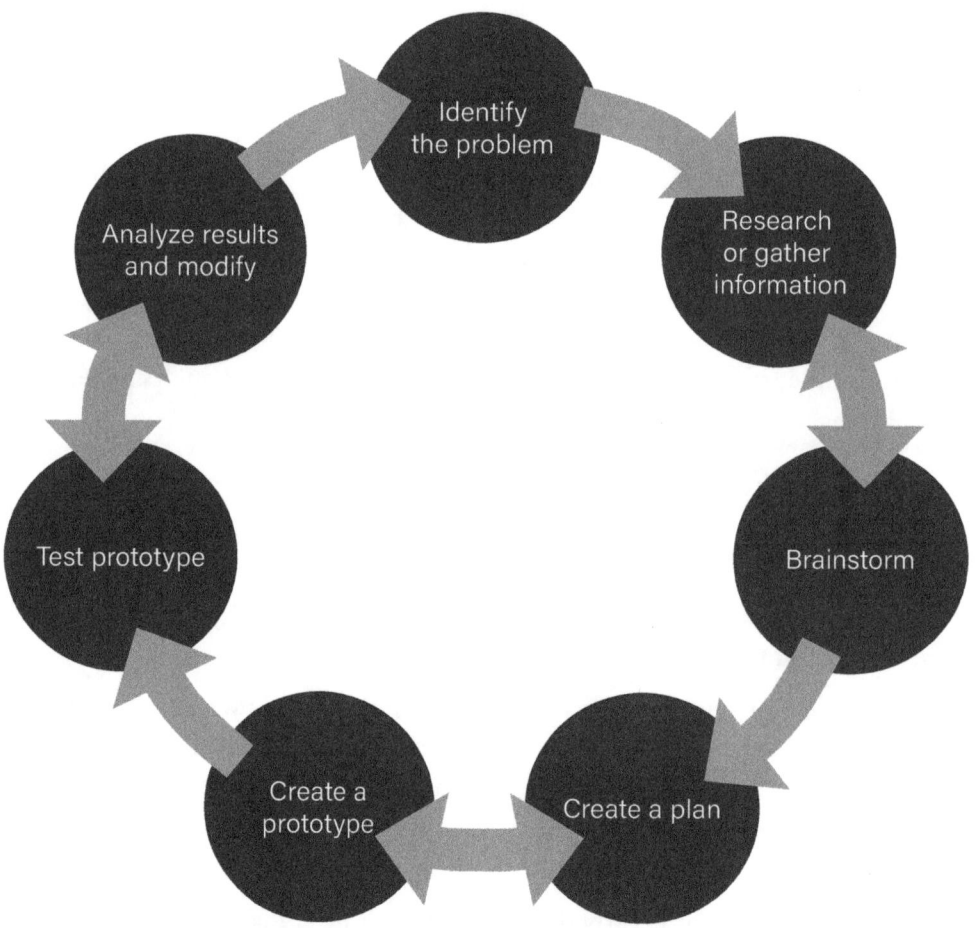

FIGURE 3.7: The engineering design process.

This planning provides the opportunity to break down the information into chunks, allowing students to begin the process of sensemaking *their* way. This is important because the brain can only, on average, handle five to seven bits of information in working memory at a time, which includes translating between languages (Sousa, 2006). *Working memory* is the small amount of information a student can hold in their mind and use in executing cognitive tasks, in contrast with *long-term memory*, which is the vast amount of information saved in one's lifetime (Cowan, 2014). Think of one bit of information being equal to one letter of the alphabet. Use the QR code to access an interactive activity that explains working memory capacity.

The more a student can make sense of the information *their* way, the more working memory is available to process other things, like translating words to their native language. You must plan in a way that allows students to make sense of what is happening before you provide context. Remember, the brain wants to make sense of the information before the details.

Working Memory Capacity
youtu.be/HJ15v6lRV6I

PRACTICE A CULTURALLY RESPONSIVE APPROACH

Learning a new language also involves learning the culture associated with the new language. The greater the social distance between the learner's native culture and the new language they are learning, the more difficulty the learner experiences in acquiring the new language (Sousa, 2006). Educators need to pay close attention to the embedded messages in their schools regarding their beliefs, values, and status pertaining to the cultures of the non-English-speaking students. All too frequently, culturally and linguistically underserved students are seen as defiant, deficient, disruptive, and disrespectful, and what they bring to the classroom is not seen as an asset but as a liability (Hollie, 2018; Sousa, 2006). Scan the QR code to access an interview I did for "Real Talk on Racism," in which I discuss issues that impact student achievement for minorities and multilingual learners.

A way to bridge the cultural divide is to highlight contributions of diverse populations to the area of study—in this case, aviation. By incorporating the contributions of diverse populations into your lesson design, you create an environment where these students feel welcome, *and* you educate native speakers on contributions of your multilingual learners' culture. This practice prepares students to live in a world that acknowledges and embraces diverse cultural and linguistic differences.

Real Talk on Racism
www.youtube.com
/watch?v=ZlP4P2Hdihw

A culturally responsive approach to this lesson is far more than just placing books around the classroom or pictures on the walls depicting cultural diversity in aviation; it requires that you gather information beforehand to inform the way that you plan and execute the lesson. When planning this lesson, you practice what you've learned about culturally responsive teaching by completing the following preparations.

- Research the contributions of diverse populations in aviation and their connection to flight that will be covered in the STEM challenge. This research uncovers two key things: (1) the aviation industry has not historically celebrated diversity, equity, and inclusion and is now taking steps to do so, and (2) minorities have made significant contributions to aviation. You collect the following resources, which will be a great addition to discussions about the forces of flight and how the following groups contributed to aviation.
    * Embracing diversity, equity, and inclusion in the aviation industry (https://calaero.edu/diversity-equity-inclusion-dei-aviation)
    * African American aviation pioneers (https://headlines.flydayton.com/10-african-american-pioneers-aviation-aerospace)
    * Latino contributions in flight (https://sandiegoairandspace.org/exhibits/online-exhibit-page/hispanic-americans-in-aviation)
    * Minority contributions in flight (https://minoritiesinaviation.aero/media)
    * Mexican Americans in Aviation (https://sandiegoairandspace.org/exhibits/online-exhibit-page/mexican-americans-in-aviation)
- Understand Maria's level of language proficiency and what she can and cannot do in reference to speaking, reading, writing, and listening according to her WIDA results (see figure 3.8).

Language Domain	Proficiency Level	Student Capabilities by Level
Reading	1.0	Students at this level generally can understand written texts that include visuals and may contain a few words or phrases in English.
Writing	1.0	Students at this level generally can communicate in writing using visuals and symbols that may contain few words in English.
Speaking	2.0	Students at this level generally can communicate ideas using words and phrases related to everyday routines or situations.
Listening	2.0	Students at this level generally can understand messages or directions involving language related to routines and familiar experiences.

FIGURE 3.8: Maria's WIDA score breakdown.

- Understanding Maria's proficiency means that you plan to:
    * Translate as necessary
    * Use visuals and gestures as much as possible
    * Use alternative words and cognates

- Remembering that Maria is from Mexico, a collectivist society, you predict that she may work best in a group setting.
- Culturally responsive teaching increases students' motivation, interest in content, and the perception of themselves as capable students, among other benefits (Will & Najarro, 2022). Teaching in a way that allows students to see themselves in the lessons by incorporating students' cultural identities and lived experiences into the classroom are tools for effective instruction and celebrating diverse cultures.

Begin the Lesson With Cultural Connection

Now that you've laid the foundation to planning, begin the lesson by making it culturally responsive to engage and interest all students.

MAKE THE CULTURAL CONNECTION TO FLIGHT

Use the following information, from the California Aeronautical University (Johnston, 2022), to introduce the topic to students.

> Diversity, equity, and inclusion are essential in every industry, and the aviation industry is no exception. Unfortunately, according to the U.S. Bureau of Labor Statistics, the recent numbers don't look good. For aircraft pilots and flight engineers in 2021, there were 5.3 percent women, 3.9 percent Black or African American, 1.5 percent Asian, and 6.1 percent Hispanic and Latino. Breaking through barriers and committing to changing the current picture of the aviation profession is a step in the right direction.

Then, provide them with the graph in figure 3.9 to demonstrate the lack of diversity in aviation (Data USA, n.d.).

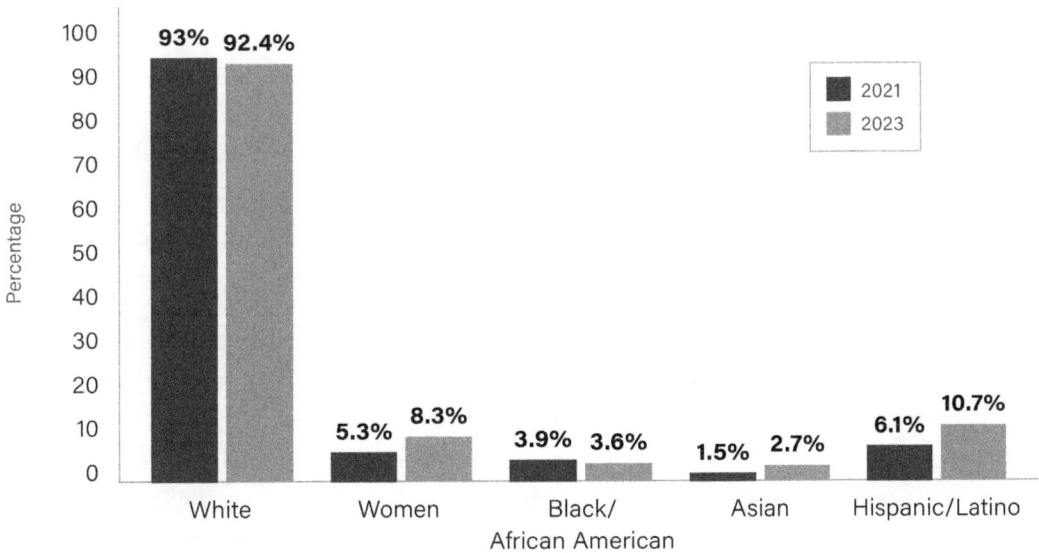

FIGURE 3.9: Lack of diversity in the aviation industry.

SUPPORT THE LESSON WITH VISUALS

Use visuals, such as a simulation, video, anchor charts, physical gestures, document cameras, digital or printed images, and illustrated vocabulary cards, to create excitement, engagement, and an opportunity for discourse. Multilingual learners can feel disoriented and anxious when learning a language in an immersive environment. Visuals support them through the process and reduce their anxiety, increase their comprehension, and even help with decoding (Southern Oklahoma State University, 2021). The use of visual aids decreases English learners' fears of giving wrong answers to questions and encourages them to engage more during lessons. When English learners feel more confident in the classroom, they are more likely to participate in the tasks and effectively absorb the new content (Halwani, 2017).

For this lesson, you decide to provide a visual of a paper airplane for students using the following activities.

- Open the lesson with a video of a person making the world record for longest paper airplane flight (youtu.be/wedcZp07raE). You ask if any student has ever made a paper airplane.

- Divide students into groups and provide each group with materials to create their own paper airplane. For students who have never made a paper airplane, designate a member of their group to show them how.

- Allow students the opportunity to fly their paper airplanes and to make notes of any adjustments they can make so it will fly better. You may need to model the note taking or create a resource for them to use.

Student actions in response to these activities function as building blocks that will provide a gateway for sensemaking on flight from vocabulary to theory. Remember, the brain wants to make sense of the information before it wants details.

DEMONSTRATE VOCABULARY

Look for cognates and visuals for essential vocabulary words students will encounter within the STEM challenge. One very effective way for students to learn vocabulary is when they encounter opportunities to explain occurrences in their own words without the vocabulary word being provided to them first. In STEM, you can use hands-on demonstrations where students describe in their own words about what they think is happening.

Use the following activity, "Paper Strip," to support students in engaging with the vocabulary terms *lift, low air pressure, high air pressure,* and *Bernoulli's Principle*.

- Conduct a hands-on demonstration of air pressure by holding a strip of paper between two fingers right below your bottom lip, then blow over the paper strip. Ask students to describe what they notice and why they think it is happening. (Teacher's note: The paper strip lifts upward. This is an example of Bernoulli's Principle, which states that faster moving air has a lower pressure than slower moving air. The air over the top, from blowing on it, is faster moving air. The air on the bottom, which is not being blown on, is slower moving air compared to the top, so the paper strip lifts up.) Scan the QR code to see an example.

- As students describe aloud what they notice (the multilingual learner may use gestures to explain their observations), write those descriptions on the board or on chart paper in the front of the room.

- Listen for key words to connect to the unit's vocabulary words. In this case, the vocabulary word is *lift*. Students may use phrases or words, such as "it's rising," "the paper is rising up," or "the paper moves upward." The multilingual learner may use a gesture—such as pointing their finger upward or demonstrating showing the paper strip in a low starting position before showing it move up—or say, "paper move" or "paper go up." The vocabulary word you are seeking to explain is *lift* or *lifting*.

Why Airplanes Fly
www.youtube.com/watch?feature=shared&v=S3cUN2B0wSA

Scan the QR code for an example of lift and other vocabulary words, including a simple explanation of how airplanes fly.

Consider the following additional activities you can use with this lesson.

- **Roto-copter (www.exploratorium.edu/explore/science/activity/roto-copter):** Use this activity to help students engage with the vocabulary terms *gravity*, *cargo*, and *load*. In this activity, students add paperclips to see how they affect the roto-copter's movement, direction, and speed.

- **Hoopster (https://annex.exploratorium.edu/science-explorer/hoopster.html):** Use this activity to explain the vocabulary words *drag* and *balance*; students use two different size paper hoops, learn how the larger hoop creates drag, and the importance of balancing the two hoops to keep the airplane leveled.

- **Spinning Blimps (https://annex.exploratorium.edu/science-explorer/spinning_blimps.html):** Use this activity to have students develop an understanding of the vocabulary words for *control*, *trials*, and *variables*, the importance of keeping track of their results, and how to control variables during experimentation.

CREATE STRATEGIC PARTNER SUPPORT GROUPS

Teachers play a key role in grouping multilingual learners and structuring activities so these students have regular opportunities to share their ideas and funds of knowledge, and to practice reading, writing, listening, and speaking in English. Grouping native or fluent speakers and multilingual learners together creates an environment where the multilingual learners are seen as competent community members and contributors in the classroom even when their English is not perfect (or, as we say in education, proficient).

One grouping arrangement, Two Pairs in a Quad, offers a strategy on teaching in a classroom with many multilingual learners at different proficiency levels while still maintaining rigor. Two Pairs in a Quad allows students to collaborate and ask questions in a small group setting as opposed to a full classroom, which can be intimidating and uncomfortable for the multilingual learner. For a deep dive into Two Pairs in a Quad, see chapter 4 (page 77).

For this activity, divide students into two pairs within a quad (four students) and instruct them to identify the parts of an airplane. If you have an odd number of students, it is OK to create a group of three or five students, but no more or less than that. This is a powerful strategy that provides the teacher with students' background knowledge of flight. Using figure 3.10, students work in pairs to identify the parts of the airplane and what they think is the function of each. This can be done in their home language.

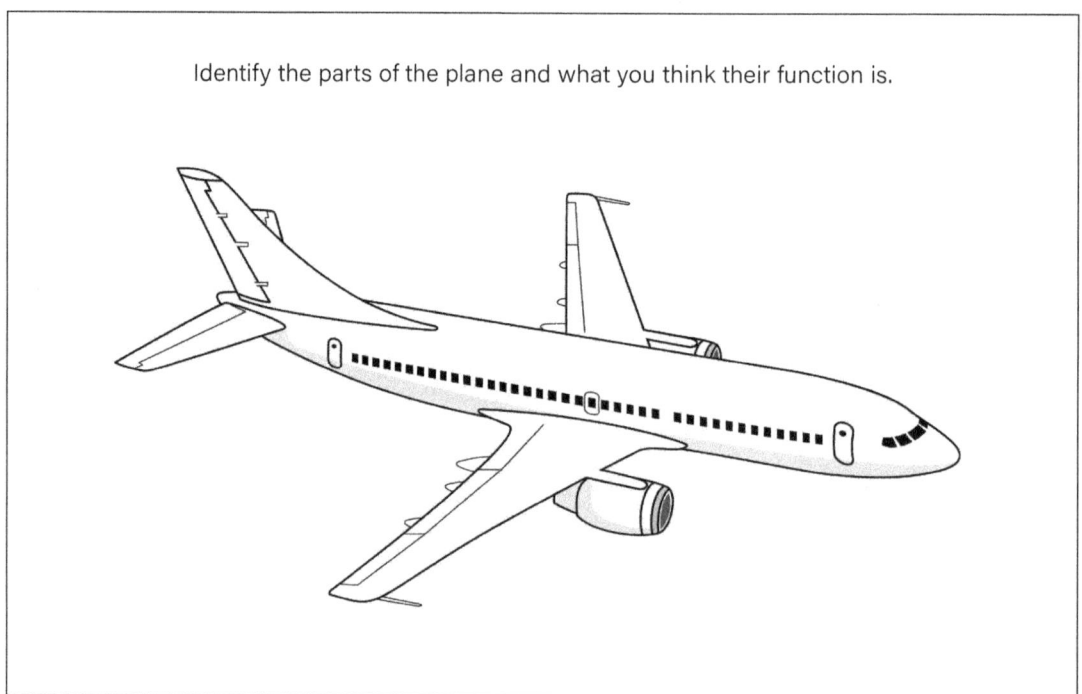

FIGURE 3.10: Blank chart for identifying parts of a plane.

The chart can be posted on the group's table or on a wall nearby. This chart remains up during the entire STEM challenge and students are required to fill in additional information throughout. This chart becomes a visual reference point for students to view as needed. The chart will become filled with new information and should become more detailed as they learn more with each lesson, as seen in figure 3.11.

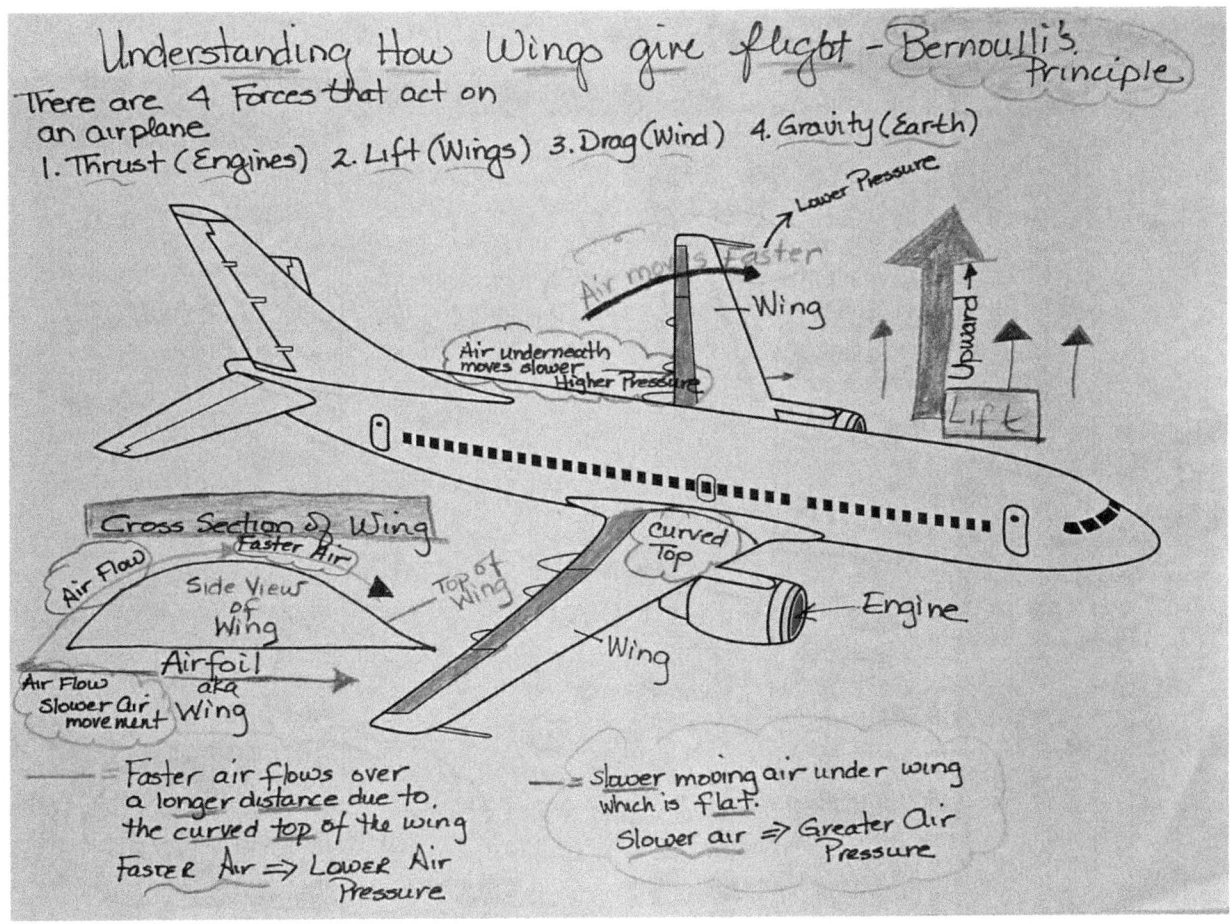

FIGURE 3.11: Completed chart for identifying parts of a plane.

Figure 3.11 shows the chart completed by the end of the unit. The chart in your classroom may have words in both English and the multilingual learner's native language. When reviewing the charts, ask the multilingual learner how to say that word in their language. If they don't know the word, you can look it up for them and instruct them to write the word on the chart next to the English words. This aligns with culturally responsive teaching principles.

CHECK FOR ALIGNMENT WITH CULTURALLY RESPONSIVE TEACHING PRINCIPLES

Finally, check how the planning aligns with the four essential components of culturally responsive teaching and the culture and language framework (Hammond, 2013; Hammond, 2015). Some of the components are items that are not evident in the planning stage, but happen in the classroom environment. The teacher-student relationship is one example. That must occur outside of lesson planning. What could you have added to the lesson to help more students connect with the content?

Table 3.5 (page 74) illustrates how the four essential components of culturally responsive teaching and the culture and language framework (Hammond, 2013; Hammond, 2015) align with this STEM challenge's aviation activities.

TABLE 3.5: Checking Alignment With Culturally Responsive Teaching

Teacher Action	Alignment
Expect the multilingual learner to contribute to the activity by ensuring that each group is including the multilingual learner. Include the multilingual learner when visiting each group and asking questions. Make eye contact with the multilingual learner so it is understood that you are including them in the conversation and using visuals and gestures during this communication.	*Culturally Responsive Teaching:* Hold high expectations for all students
If applicable, allow the use of students' home language when two multilingual learners speak the same language. Allow the multilingual learners to work collaboratively on the assignment.	*Culturally Responsive Teaching:* Appreciate different communication styles
Create a safe environment.	*Culture and Language Framework:* Create a community of learners and safe learning environment
Allow students to use gestures or visuals to represent their thoughts.	*Culture and Language Framework:* Understand the levels of language acquisition

Key Takeaways

Teaching in a deliberate manner that makes content relevant to students helps them succeed both in terms of quantitative measures, such as high test scores, and qualitative measures, such as becoming lifelong learners able to ask critical questions about the world around them, both in and out of school (Aronson & Laughter, 2016). The design and implementation of a STEM challenge must include the needs of *all* students. The following points can help create an inclusive environment.

- Culturally responsive teaching connects students' cultures, languages, and life experiences with classroom learning. These connections help students access rigorous curriculum and develop higher-level academic skills.

- Culturally responsive teaching informs your practice so you can make better teaching choices for eliciting, engaging, motivating, supporting, and expanding the intellectual capacity of *all* your students (Hammond, 2015).

- Culturally responsive teaching is one of your most powerful tools for helping your students find their way out of the achievement gap that impacts underserved multilingual learners, students in low-income households, and students of color.

- All human beings have biases, whether implicit or explicit. These biases can skew our perceptions of our students and have grave impacts on the expectations we have for diverse students and the rigor of their work.

- Understanding and acknowledging your biases and recognizing cognitive dissonance will help you better teach your students and set rigorous goals for them to achieve.

- Knowing your students can help you adapt your instruction so that all students can feel welcomed and appreciated.

- You don't need to learn about all the customs, foods, and beliefs of every student in your class. You can look for patterns, also known as cultural archetypes, to guide you.

- To be an effective teacher of multilingual learners as a general education teacher, you must collaborate with your team teachers, guidance counselors, and the parents to obtain as complete a picture as possible of the student. Team plan with the English language instruction or bilingual teacher and reach out to the district's bilingual supervisors to discuss your ideas and concerns, or to garner advice on how to amend your lessons.

- Background information on the student is essential to creating a safe environment, creating relationships, and most of all, knowing your English learner.

- Self-reflection is paramount to becoming an effective culturally responsive teacher and identifying your blind spots.

- Designing lessons in STEM courses with an awareness of individualism and collectivism involves tailoring the educational experience to support diverse cultural orientations. This can enhance student engagement, foster deeper understanding, and promote inclusivity in the classroom.

CHAPTER 4

Using Collaborative Learning Groups to Support Language Acquisition and Sustain Rigor

> *Language is a product of interaction and learning, not a precursor or prerequisite.*
>
> **—NATIONAL ACADEMY OF SCIENCES**

Cooperative learning groups are an effective strategy teachers can use to support multilingual learners' unique needs while sustaining rigor for all students. It's important to know that by "grouping," I do *not* mean grouping students according to academic ability—what some districts may call *tracking*, *phasing*, or *streaming*. These terms refer to a system in which students are divided into classes based on their overall academic achievement (Barrington, 2020). *Grouping* in the context of this chapter simply means placing the students in your classroom into cooperative learning groups that have a mixture of academic ability and language proficiency. This does not mean placing all multilingual learners together. That defies the entire purpose of using groupings to build language and content simultaneously.

I saw the need for grouping students during my work as an assistant principal. One day when I visited a mathematics classroom, I immediately noticed that the multilingual learners were not sitting together but were spread far apart from each other in the back of the classroom. It was clear they were not engaged in the lesson. I asked them, one at a time, what they were learning that day. Neither of them could tell me. Each student's language proficiency was around levels 1 and 2 (entering and beginning), yet these multilingual learners hadn't received language or academic supports. At the end of class, I asked the teacher why her multilingual learners were spread out in the back of the classroom. Her response was one I heard from many general education teachers: "I don't understand what they're saying, so I separated them because I didn't know if they were speaking about the class or just talking in class." Though I understood the logic, by separating these students, the teacher took away one of the strongest supports multilingual learners have—their language. This meant they could not help each other by translating what the teacher was saying or explaining the content in a language they understood. I asked her how they were doing in the class, and she confirmed they were failing. The teacher informed me that she did not know what to do nor how to teach them; she had never received direction or assistance from administration about multilingual learners' needs and strategies for meeting them. Imagine how the students' experience in that class might have been different if the teacher knew how to place them in cooperative learning groups where they could pool their resources, learn from one another, maintain access to their language, and practice using English in a safe environment.

Placing students into groups is important. Research shows that students need to engage in productive discourse and interactions with others so they can hear the language in multiple registers (Francis & Stephens, 2018). *Registers* refers to the way a person uses language in different circumstances like whole group, small group, pair work, and individual thinking time. Grouping students allows for the seamless differentiation of instruction that does not reduce the rigor of the lesson, but instead provides multilingual learners, struggling native speakers, and advanced students (whether native speakers or multilingual learners) to work at their zone of proximal development. The *zone of proximal* development is defined as the distance or the cognitive gap between what a child can do unaided and what the child can do in coordination with a more skilled person (Gibbons, 2015).

In this chapter, I provide a foundational understanding of cooperative learning models, focusing specifically on a heterogeneous model I call *Two Pairs in a Quad*. This method is a powerful tool that allows multilingual learners to work in diverse group configurations that maximize exposure to linguistic and academic content. With the help of fictional students Jean Pierre Léger, Linh Kha, and Amihan Villanueva, I demonstrate how to create Two Pairs in a Quad groupings and use them to facilitate students' collaborative work during lessons. Toward the end of the chapter, I provide a STEM challenge that incorporates Two Pairs in a Quad as well as scaffolds for multilingual learners.

Cooperative Learning Models

Student grouping, also known as cooperative learning, is most powerful with practicing and deepening new knowledge because it gets the information to long-term memory. For more on this topic, use the QR code to view "The Journey: Memory and the Hippocampus."

The Journey: Memory and the Hippocampus

https://youtu.be/CCAZz4X kW6s?si=weUSSxRrjaHW_Y5d

Successful cooperative learning acknowledges that students learn differently, and grouping enables the teacher to strategically meet the needs of *all* students. Successful cooperative learning increases test scores and engagement in the classroom (Dotson, 2001). It also teaches all students, regardless of home language, crucial social skills like interacting with others in an appropriate way, especially as globalization and advances in technology continue to increase the quantity of accessible information and the need for collaboration (Willis, 2021). Neurologist and educator Judy Willis is an authority on learning and the brain, correlating research on the subject with best teaching practices. In the Edutopia article, "How Cooperative Learning Can Benefit Students This Year," Willis (2021) writes that the COVID-19 pandemic led to the decrease in interpersonal contact and collaboration between students because students spent more time in the digital world than interacting with peers. Cooperative learning can guide the brain's reconstruction and boost social and emotional cue awareness, which was minimized during remote learning.

Although research has shown cooperative learning is effective in student learning, engagement, and achievement, there are still discussions and concerns about whether it is beneficial for gifted or advanced students. The next section addresses some of the concerns educators and parents have regarding cooperative learning.

Concerns About Cooperative Learning

Educators, parents, and students are sometimes skeptical about cooperative learning. When asked about their feelings and concerns, students responded in the following ways (Middlecamp, n.d.):

- People need to go at different speeds.
- Someone may try to take over the group.
- Quiet people may not feel comfortable.
- Sometimes people just don't get along.
- People may not pull their weight.
- It is not fair!
- A concept may not be understood as well if a person does not have to figure it out.
- The time spent talking about irrelevant topics is unbelievable.

Teachers share these concerns too. They're often reluctant to create cooperative learning groups in the classroom, afraid that the classroom will be too noisy, one student will be doing all of the work, students may belittle or ignore groupmates that they perceive to be low achievers, or that some students may find cooperative learning time to be party time as opposed to learning time (Slavin, 2014).

These concerns are valid when cooperative learning is not well organized. However, learning groups like Two Pairs in a Quad greatly reduce these concerns. When students are in well-designed teams, they easily shift between direct instruction with the teacher and pair work with their face partner or shoulder partner without the interruption of rearranging seats and desks, which wastes instructional time. Working in quads, students become supportive of each other and learn how to assist peers to form a strong team bond. The quad becomes a safe place for multilingual learners to take chances both academically and linguistically. Additionally, students aren't limited to solely working in their quads. Teachers can (and should) create activities and lessons where students can intermingle with their classmates, which creates additional opportunities for multilingual learners to practice the language and share their ideas while contributing to the lesson topic discussions.

Another common concern about cooperative learning groups is that they reduce rigor for high-achieving students, but this isn't true. Research shows that the learning progress of high-achieving (the academically top 33 percent) or gifted students (the top 5 percent) is not negatively impacted when working in heterogeneous cooperative groups, although there are times when high-ability students should work in isolation from other students, and there are times when gifted students should compete to see who is best (Johnson & Johnson, 2019). Professors David W. Johnson and Roger T. Johnson (1992) conducted nine studies over a fifteen-year period to examine the impact of high-ability students learning "individually, competitively, cooperatively in homogeneously high-ability groups, and cooperatively in academically heterogeneous groups." Consider the following findings.

- High-ability groups do benefit academically from cooperative learning groups due to the exchange of ideas within the group. This exchange of ideas occurs when cooperative groups are carefully structured.

- Lower-achieving peers do not decrease the critical thinking and higher-level reasoning of high-achieving students, which is a common belief. In fact, the opposite occurs. Cooperative learning groups develop all students' higher-order thinking skills because the students have different points of view that can enhance conversations because they can challenge each other's assumptions and bring different information to the argument.

In addition to these benefits, cooperative groups provide gifted students with opportunities to build social skills, an area some gifted students struggle with (Ferlazzo, 2019; Kagan, 2015a).

To ensure that students receive the benefits of cooperative learning groups and avoid the pitfalls, teachers should adopt psychologist Spencer Kagan's (n.d.) four basic principles of cooperative learning that are symbolized by the acronym PIES. The four principles are as follows.

- **Positive interdependence:** A gain for one must be a gain for others. Instead of having students raise their hands to answer a question, have students pair up and take turns sharing their answers with each other. This allows all students to have a voice in the answer and there are no winners and losers as occurs when students raise their hands, and the teacher selects a student to answer (hence the winner). Not only does this motivate all students to participate, but also provides consistent opportunities for multilingual learners to hear and practice the target language.

- **Individual accountability:** Students are responsible for their performance and held accountable for participating in the discussion and lesson. In quads, this means that all members are held accountable for interacting with their partners by completing their part of the task.

- **Equal participation:** The conventional classroom setup epitomizes unequal involvement. Typically, when a teacher poses a question, only a handful of eager volunteers vie for the chance to answer. Who usually steps forward? The high-performing students. Consequently, classroom dynamics often revolve around dialogue primarily between the teacher and these high achievers, neglecting the participation of those who may benefit most from engagement. In the ideal classroom, all students should participate equally, even the unmotivated students who seldom raise their hands to participate. By implementing Two Pairs in a Quad, the students in the quad are specifically selected, as are the activities, so that each student equally participates. This improves self-esteem and indirectly shows all students that they are important as well as what they have to say.

- **Simultaneous interaction:** Placing students in groups allows for all students to engage with the lesson simultaneously, in contrast to whole-group instruction where only one student may interact with the teacher's instruction. The teacher calls on one student at a time in nongroup classrooms. But in quads, students are afforded the opportunity to speak with their shoulder partner, face partner, and entire quad, switching roles between listener and speaker.

These principles distinguish cooperative learning from other forms of learning and are fundamental for its success (Kagan, n.d.). For a more in-depth review of these principles, visit www.kaganonline.com. When students are in cooperative learning groups and the PIES principles are instilled in the instructional framework, participation goes up dramatically (Kagan, n.d.). The following section discusses additional benefits of cooperative learning.

Benefits of Cooperative Learning

Cooperative learning benefits all students, not just multilingual learners. Consider the following benefits of cooperative learning.

- Kagan recommends that students get the chance to interact with each other and the information after ten minutes of instruction (Kagan, 2015a). This approach breaks the content into digestible chunks. After students share with each other, the teacher should check for understanding by determining a method for sharing, such as calling on a random group, instructing students to hold up dry erase boards with their responses, or something else. Having a variety of grouping strategies will help keep things new and novel in the classroom (Ward, 1987).

- As students accumulate more favorable experiences within their groups, they grow increasingly at ease with participation and academic daring, such as being open to the possibility of being incorrect, suggesting ideas, defending their viewpoints, and similar actions (Willis, 2021).

- Given the impracticality of providing individualized teacher attention to all students regularly throughout the day, cooperative learning groups offer a solution by diminishing student reliance on teachers for guidance, behavior management, and feedback on progress (Willis, 2021).

- The collaborative nature of group interdependence in cooperative learning enhances emotional sensitivity and communication abilities. Through the planning involved in cooperative learning, students are entrusted with decision-making and conflict-resolution responsibilities. In times of flux and uncertainty, the structured and growth-oriented experiences of well-organized cooperative learning provide a reassuring anchor (Willis, 2021).

Pause for a moment to reflect on what you've read. What connections can you make between effective strategies in teaching multilingual learners and cooperative learning groups? How can you envision using them in your classroom?

Types of Cooperative Learning Models

Educators often question how to group students when considering cooperative learning models. Although there are a variety of options, I will focus on three popular ways of grouping students for cooperative learning: (1) mixed group, (2) clustered group, and (3) grouping by proficiency level. A *mixed group* is where multilingual learners are grouped together at various levels of second language development. Some districts favor this model because it provides multilingual learners with different language models to support second language acquisition. A *clustered group* organizes students based on the similarities of their English-language ability. This model allows teachers to target language instruction specifically to their common level of language development (Mora-Flores, 2011).

Both mixed and clustered grouping have their advantages and disadvantages. Having a mixed group provides the multilingual learner with opportunities to learn new discourse patterns and new ways to interact using language (Francis & Stephens, 2018). Mixed groups are also beneficial because they provide an opportunity for native English speakers to recognize the multilingual learners' contributions to the lesson, and the native speaker will begin to engage more with the multilingual learner in peer-to-peer discussions. This contributes to helping the multilingual learner feel part of the classroom community. Having multilingual learners clustered together can help the teacher target language instruction specifically to the needs of students at their common level of English language development and the teacher can use the native language as a scaffold to accelerate multilingual learners' academic language development (de Jong & Commins, n.d.; Mora-Flores, 2011). However, mixed groups can decrease rigor and engagement for the high-performing students if the work is not differentiated or if the teacher only teaches to the lower or middle group.

Multilingual learners also may not have the opportunity to hear the language in its context nor the ability or time to practice, which may discourage use of the target language.

To learn more about how the brain is wired for relationships, scan the QR code to my YouTube video.

Research shows that grouping depends on a variety of factors and decisions, such as multilingual learners' demographics and proficiency levels, teacher capacity, and the trade-offs that occur when choosing one grouping model over another (Najarro, 2023). Regardless of the selected model, grouping does benefit multilingual learners in some capacity versus not using a grouping model at all. If a teacher uses a question-and-answer format with the whole class in place of a grouping model, chances are great that multilingual learners will not risk embarrassment by speaking in front of the whole class, opting not to participate even if they know the answer (Kagan & High, 2002). The student will lose interest in the lesson and lose new content learning, not to mention missing opportunities to practice the language with peers. In contrast, if the teacher uses a grouping model, multilingual learners have the chance to operate in a very small environment where they solely interact with a limited number of partners. In this dynamic, multilingual learners can more easily engage in the lesson, take risks, and participate because they have a support system within the group and receive encouragement from their partners, which aids students in learning both content and language.

Student-to-Student Relationships
https://youtu.be/HN2jpQIAHGg

Consider the following grouping models, which fall into the third category based on proficiency levels.

- **Homogeneous grouping:** Group together students with similar proficiency levels. This allows targeted instruction to cater to their specific needs.

- **Heterogeneous grouping:** Mix students with different proficiency levels. This allows more proficient students to serve as models and support for those with less proficiency.

- **Buddy system:** Pair a lower-proficiency student with a higher-proficiency student. The higher-proficiency student can help clarify concepts and provide additional support.

- **Interest groups:** Group students based on shared interests or topics. This can motivate multilingual students to participate more, as they're discussing something they're passionate about.

- **Task-based grouping:** Design activities where each group member has a specific role that suits their proficiency level. For instance, a beginner might be responsible for drawing while an intermediate student writes a description.

- **Rotating groups:** Regularly change the grouping. This ensures that multilingual learners get a chance to interact with different peers and benefit from varied dynamics.

- **Skill-specific groups:** Sometimes it's beneficial to group students based on a particular language skill they need to improve, such as reading, writing, speaking, or listening.
- **Cultural grouping:** Occasionally, it might be helpful to group students based on shared cultural backgrounds, especially if cultural context is crucial for a particular lesson or activity.
- **Mixed-age grouping:** In some cases, older students with more proficiency can mentor younger or less proficient students.
- **Whole-class activities:** Do not always rely on group activities. Whole-class instruction and activities can also be beneficial, especially when introducing new concepts or when the lesson is designed to be accessible for all proficiency levels.

Of these, heterogeneous teams are particularly effective; this is the method adopted by Spencer Kagan, educational psychologist and founder of Kagan Cooperative Learning. Kagan developed a method of cooperative learning that focuses on student engagement in the classroom by dividing students into groups of four, called *quads*. Within the quads, students are paired into shoulder partners and face partners; Kagan calls these pairs heterogeneous teams or mixed teams. The teams consist of students labeled as H for high achiever, HM for high medium achiever, LM for low medium achiever, and L for low achiever. That means groups comprise diverse genders, races, and ability levels (Kagan, 2015b).

To learn more about Kagan cooperative learning structures, go to www.kaganonline.com.

Two Pairs in a Quad

This book focuses on Kagan's heterogeneous model using a grouping style I call Two Pairs in a Quad, which is a mixture of several grouping styles mentioned in the previous section. Two Pairs in a Quad is a cooperative learning group where students have the flexibility to work individually, in diverse pairs, or as a whole quad, depending on the assignment. Two Pairs in a Quad allows native English speakers and multilingual learners to work together while maintaining the rigor of the lesson so that each student works within their zone of proximal development. This way, the teacher can avoid teaching to the middle, which is not beneficial to high- nor low-level students. Two Pairs in a Quad is beneficial for multilingual learners because it creates an environment where they can comfortably interact with a small number of peers, take risks, practice the new language with confidence, listen to the content language in its true context, and experience psychological safety. Teachers can select group members by ability level, language proficiency, compatibility, and content knowledge strengths.

Figure 4.1 illustrates how the quads are structured, using face partners (A1 and A2; B1 and B2) and shoulder partners (A1 and B1; A2 and B2).

FIGURE 4.1: Quad structure showing shoulder partners and face partners.

Similar to Kagan's model, each quad contains students with the following labels: high (H), medium high (MH), medium low (ML), and low (L). Note these designations are for the teacher only; they are not known to students. Preferably one student should speak the multilingual learner's language, but all four group members need to be compatible. The MH student should have strong content knowledge because they will be the H student's face partner. The ML student must be academically stronger than the L student and, if possible, speak the multilingual learner's language. Remember, the multilingual learner does not have to be the L student because multilingual learners may have strong content knowledge even with low English proficiency. In this case, the multilingual learner may be the ML student based on their content knowledge, and the native speaker may be the L student.

It's important to note that these designations are not meant as static labels for students; they are flexible designations that allow the teacher to meet each student's academic and social needs. The teacher aims to maintain rigor and ensure that students who are advanced academically are appropriately challenged rather than ignored while students in need of more support receive the help they need. And a student's designation may change based on the assignment or activity—a student who is high in one activity or content area may be low in another and vice versa. The teacher should evaluate students' ability levels each time they assign quads to ensure that all students receive the level of challenge or support they need to engage with the activity.

So, how should students be paired as shoulder partners and face partners? Form face partners and shoulder partners as follows.

- **Face partners:** H student paired with MH student and ML student paired with L student
- **Shoulder partners:** H student paired with ML student and MH student paired with L student

Figure 4.2 illustrates these pairings.

FIGURE 4.2: Face partners and shoulder partners based on student designations.

These multiple pairings within the quad provide students several avenues of support in one location before resorting to teacher intervention. Students in this setting can listen to multiple points of view, get assistance from peers, clarify misconceptions and incorrect information—all in a small setting that reduces the embarrassment or discomfort they might experience in the whole-class dynamic. Ultimately, this grouping style benefits language development and learning by providing students with a safe mini environment where they can practice and use language in an authentic way. It is also valuable for native speakers and is an effective method of teaching when the classroom contains multilingual learners at multiple levels of language proficiency as well as native speakers. Kagan (2015a) said it best:

> Cooperative learning is an educator's dream: It gives an incredible amount of leverage because when the leverage is in the right place, the students can obtain a mechanical advantage and can lift a large load with little effort and a wide range of positive outcomes. (pp. 4.1–4.2).

It's important to note that this type of cooperative learning group differs from group work. They are not the same thing! In group work, students need each other to complete a task with each student having a specific role or responsibility within the group. The focus of group work is often on completing the task, and individual accountability may vary depending on the group-work structure. Note that although both group work and collaborative learning groups involve students working together, collaborative learning using Two Pairs in a Quad places a stronger emphasis on the *process of learning* with mutual support from the quad members and a shared responsibility for the contributions each person provides in the learning process. Two Pairs in a Quad provides students the opportunity to learn language and content simultaneously. Table 4.1 illustrates how group work and collaborative learning groups like Two Pairs in a Quad are similar and different.

TABLE 4.1: Similarities and Differences Between Group Work and Two Pairs in a Quad

Group Work	Two Pairs in a Quad
Students collaborate, interacting with each other to achieve a common goal or complete a task.	Students interact with each other, but their interactions are highly structured. Students work with specific people in their group. Students work first with their face partner. If a student is having difficulty, they then refer to their shoulder partner. If the student still needs assistance, they ask the question to the entire quad. If that does not provide a solution, they may call the teacher over for assistance.
Each member's unique skills, perspectives, and strengths become integrated to enhance productivity and creativity.	Each student's unique skills, perspectives, and strengths are available to help group members as needed, but each student completes work at their individual level.
Effective group work relies on clear communication, mutual respect, and a shared understanding of objectives.	Effective work within a quad relies on clear communication, mutual respect, and a shared understanding of objectives.

Group Work	Two Pairs in a Quad
Group work includes planning, division of responsibilities, regular meetings for progress updates, and collective problem solving.	Each student understands how the quad works and follows the rules for interacting with other members: face to face first, shoulder to shoulder second, then the entire quad, leaving teacher intervention as the last option.
Success hinges on members' ability to work cooperatively, manage conflicts constructively, and contribute equally toward the achievement of the group's goals.	Success hinges on members' ability to work cooperatively, manage conflicts constructively, and contribute equally toward the achievement of the group's goals.

Cooperative learning presents distinct advantages for both English learners and native English speakers compared to working in a group of four individuals. For English learners, cooperative learning provides valuable opportunities for language practice and acquisition in authentic, interactive contexts. Engaging with peers allows them to receive immediate feedback, clarify misunderstandings, and gain confidence in expressing themselves in English. Additionally, cooperative tasks encourage collaboration and mutual support, enabling English learners to develop language skills while also enhancing their social and cognitive abilities. For native English speakers, cooperative learning offers the chance to strengthen communication skills through meaningful interactions with diverse language learners. By collaborating with peers who are learning English, native speakers can refine their articulation, empathy, and cultural awareness while also reinforcing their understanding of course content through teaching and explanation. In summary, cooperative learning benefits both English learners and native English speakers by fostering language development, intercultural competence, and academic achievement within a supportive and inclusive learning environment. Before we practice creating Two Pairs in a Quad, it's time to meet your new group of multilingual learners for this chapter.

Meet Jean Pierre, Linh, and Amihan

Jean Pierre Léger is from Haiti and speaks both Haitian-Creole and English. Linh Kha was born in the United States; she lived in Little Saigon, attended a Vietnamese school until grade 6, and speaks primarily Vietnamese with some English. Amihan Villanueva was born in the United States, but she only speaks her home language of Tagalog from the Philippines. Her family enrolled her in a Filipino school for elementary school and sixth grade. This is her first time in an American school.

Figure 4.3 (page 89) contains Jean Pierre's bio and academic profile. Scan the QR code to view a message from Jean Pierre.

Meet Jean Pierre Léger
https://youtu.be/OW-KCu6_emg

Jean Pierre Léger

Jean Pierre Léger was born an only child in Gonaïves, a city in northern Haiti. Gonaïves is an agricultural municipality. Cotton, sugar, coffee, mango, banana, and cabinet wood are among its most common exports. In Haiti, there are two classes of people: a tiny cluster of the wealthy and everyone else. Most Haitians live on two dollars a day or less. Jean Pierre was fortunate to be born into that tiny cluster of wealth. His father grew up without much money and without electricity or running water. Jean Pierre's grandparents, who were illiterate, somehow were able to pay for twelve years of schooling for his father. His father was also one of the few who were awarded a college scholarship. With this college education, Jean Pierre's father was able to start a business where he sells imported consumer electronics and computer equipment and provides internet and other computer-related services. His father is one of Haiti's few stories of a person who grew up in a low-income neighborhood, earned a college degree within the country, and was able to have access to a career instead of the low-wage employment that is quite common in Haiti.

Jean Pierre was thirteen years old when he emigrated with his father from Haiti to the United States. Before Jean Pierre and his father moved, his father had sold his business interests in Haiti and invested in similar interests in the United States.

Jean Pierre is an avid guitar player, an instrument he learned to play from his mother, who died when he was eight years old. Jean Pierre wants to continue to play in his American school and hopes to be an engineer.

Name	Country of Birth	Primary Language	Level of Language Proficiency	Age	Grade
Jean Pierre	Haiti	Biliterate: Haitian-Creole and English	Level 4: Expanding	Thirteen	Seventh
WIDA Results					
Listening: 6.0		Speaking: 3.9		Reading: 3.3	Writing: 4.2

Student Capabilities By Level
Listening (proficiency level 6):
Students can understand oral language in English and participate in all academic classes. This includes synthesizing information from multiple speakers, creating models or visuals to represent detailed information presented orally, recognizing language that conveys information with precision and accuracy, and identifying strengths and limitations of different points of view.
Speaking (proficiency level 3):
Students can communicate ideas and details orally in English using several connected sentences and can participate in short conversations in school. This includes relating stories or events, sharing ideas, and providing details; describing processes or procedures; and giving opinions with reasons.
Reading (proficiency level 3):
Students can understand written language related to common topics in school and can participate in class discussions. This includes classifying main ideas and examples in written information; identifying steps in written processes and procedures; identifying main information that tells who, what, when, or where something happened; and recognizing language related to claims and supporting evidence.
Writing (proficiency level 4):
Students can communicate in writing in English using language related to specific topics in school. This includes producing papers describing specific ideas or concepts; creating explanatory text that includes details or examples; narrating stories with details of people, events, and situations; and providing opinions supported by reasons with details.

FIGURE 4.3: Jean Pierre Léger's bio and academic profile.

Meet Linh Kha
https://youtu.be/AjaPm9bjh-w

Figure 4.4 contains Linh Kha's bio and academic profile. Scan the QR code to view a message from Linh Kha.

Linh Kha

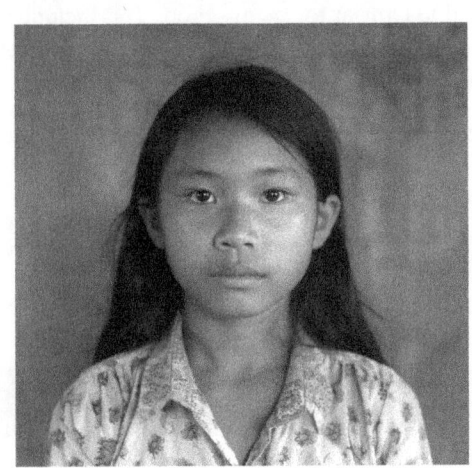

Linh Kha, an only child, was born in the United States in the Clarendon neighborhood of Arlington, Virginia. Clarendon is also called "Little Saigon." This neighborhood became the hub of Vietnamese commerce and social activity during the late 1970s and early 1980s. Linh speaks fluent Vietnamese. Her parents do not speak any English. Linh spent up to sixth grade in a Vietnamese school in Virginia, where she was only exposed to Vietnamese. This is Linh's first year in an English-speaking school.

Linh is very quiet and shy, and she excelled at mathematics at her Vietnamese school, where she was on the honor roll and helped other students in mathematics in the school's after-school program.

Linh hopes one day to become a research scientist and find a cure for lupus, a disease her grandmother had, whom she was very close to when she was growing up. Finding a cure for diseases to help other people is very important to her.

Name	Country of Birth	Primary Language	Level of Language Proficiency	Age	Grade
Linh Kha	United States	Vietnamese	Level 2: Beginning	Thirteen	Seventh
WIDA Results					
Listening: 4.9		Speaking: 1.8	Reading: 1.7		Writing: 2.3
Student Capabilities By Level					

Listening (proficiency level 4):

Students can understand oral language in English related to specific topics in school and can participate in class discussions. This includes exchanging information and ideas with others, applying key information about processes or concepts presented orally, and identifying points of view or positions on issues in oral discussions.

Speaking (proficiency level 1):

Students can communicate orally in English using gestures and language may contain a few words. This includes asking and answering simple questions about what, when, or where something happened; naming familiar objects, people, and pictures; showing how to solve problems using words and gestures; and expressing personal preferences.

Reading (proficiency level 1):
Students can understand written texts that include visuals and may contain a few words or phrases in English. This includes interpreting information from graphs, charts, and other visual information; identifying steps in processes presented in graphs or short texts with illustrations; identifying words and phrases that express opinions and claims; and comprehending short text with illustrations and simple and familiar language.
Writing (proficiency level 2):
Students can communicate in written English using language related to familiar topics in school. This includes describing ideas or concepts using phrases or short sentences; stating steps in processes or procedures; labeling illustrations describing what, when, or where something happened; and expressing opinions about specific topics or situations.

FIGURE 4.4: Linh Kha's bio and academic profile.

Figure 4.5 contains Amihan Villanueva's bio and academic profile. Scan the QR code on page 93 to view a message from Amihan.

Amihan Villanueva

Amihan Villanueva was born in San Diego, California, in the United States. She is an outgoing teenager who is very involved in her Filipino heritage. Amihan hopes one day to be an engineer. Her parents decided to put Amihan in a Filipino school because of the issues that occurred with her older brother. He was embarrassed of being Filipino and did not want anything to do with the culture. He only wanted to assimilate with American culture. Amihan's parents—wanting to ensure that what happened with her brother did not happen to her, and for her to be proud and connected to her Filipino culture and heritage—enrolled her in a Filipino school in San Diego at the age of six, where she learned Filipino culture and history in addition to the Tagalog language until the age of twelve, when her parents enrolled her in an American school. When Amihan entered the American school, she could not read, write, nor speak any English. Amihan likes to read about science and do experiments that she creates on her own with items around the house. She's an avid reader and LEGO enthusiast. Her father has given her LEGO sets since early childhood and she has continued to collect them and build structures she imagines from the many space exploration books she reads from the local library. Amihan loves to help others and does not hesitate to get involved in school functions.

Name	Country of Birth	Primary Language	Level of Language Proficiency	Age	Grade
Amihan Villanueva	United States	Tagalog from Philippines	Level 1: Entering	Thirteen	Seventh
WIDA Results					
Listening: 1.7		Speaking: 1.4	Reading: 1.7		Writing: 1.1
Student Capabilities By Level					
Listening (proficiency level 1): Students can understand oral messages that include visuals and gestures and may contain a few everyday words or phrases in English. This includes recognizing familiar words and phrases in conversations; following one-step oral directions; matching information from oral descriptions to objects, figures, or illustrations; and showing agreement or disagreement with oral statements.					
Speaking (proficiency level 1): Students can communicate orally in English using gestures and language may contain a few words. This includes asking and answering simple questions about what, when, or where something happened; naming familiar objects, people, and pictures; showing how to solve problems using words and gestures; and expressing personal preferences.					
Reading (proficiency level 1): Students can understand written texts that includes visuals and may contain a few words or phrases in English. This includes interpreting information from graphs, charts, and other visual information; identifying steps in processes presented in graphs or short texts with illustrations; identifying words and phrases that express opinions and claims; and comprehending short text with illustrations and simple, familiar language.					
Writing (proficiency level 1): Students can communicate in written English using language related to familiar topics in school. This includes describing ideas or concepts using texts and illustrations, labeling steps in processes presented in graphs or short texts, stating opinions or preferences through text and illustrations, and sharing personal experiences through drawings and words.					

FIGURE 4.5: Amihan Villanueva's bio and academic profile.

Collaborative Learning Groups in Practice

Now that you've met your students, it's time to assign them to their quads. The following section describes how to do this.

Meet Amihan
https://youtu.be/0tKOmxIzaHE

How to Create Two Pairs in a Quad

You learned in chapter 3 (page 41) that teachers should consider multilingual learners' range of experiences, languages, cultures, and funds of knowledge to meet their social, emotional, and academic needs. These same considerations are needed when creating the Two Pairs in a Quad groupings. Consider the following questions when creating these quads (Morgan, n.d.).

- "How can I place multilingual learners where they will receive the most equitable access to learning and quality of service?"
- "What will provide the most supportive learning environment?"

There are several steps to creating effective Two Pairs in a Quad groupings. The following sections discuss these in more detail.

KNOW YOUR STUDENTS

The first step to creating successful groupings is to know your students; this means knowing the following information about each student. While you may use other characteristics in determining your quads, each quad *must* have a top 25 percent academic student.

- Level of language acquisition
- English language proficiency scores in reading, writing, speaking, and listening
- The students' personality and who their friends are among classroom peers
- Academic strengths and weaknesses
- The academic top 25 percent of the class; each quad will need one member from this group
- Which, if any, classroom peers can speak the multilingual learner's language

After considering these factors, you create your quads. For the Jean Pierre Léger, Linh Kha, Bremen Robinson, and Lindsey Pearson group, the quad includes two multilingual learners (Jean Pierre and Linh Kha) and two native English speakers (Bremen Robinson and Lindsey Pearson). For the second quad, you create a quad with three native English speakers and Amihan Villanueva, a multilingual learner. You use the following key details about each student to make your determination (see table 4.2). The remaining students in your class are all native English speakers and you follow the same process in determining their quads.

TABLE 4.2: Explanation of Quads 1 and 2

Quad 1	**Bremen Robinson:** Bremen is a native English speaker. He is in the top 25 percent of the class academically and is very strong in science.
	Lindsey Pearson: Lindsey is a native English speaker. She is academically strong in science, works well with everyone, loves learning languages, and is eager to try speaking other languages so that her multilingual peers feel comfortable. She is also a highly empathetic student.
	Jean Pierre Léger: Jean Pierre is a multilingual learner. Although he is new to the United States, he is biliterate. His overall language acquisition is at a level 4 and his high WIDA scores enable him to communicate with Linh Kha, providing details and sharing ideas.
	Linh Kha: Linh Kha is a multilingual learner. Her overall language acquisition proficiency is at a level 2. Her WIDA scores indicate that she can identify points of view in oral discussions, but she needs visuals, short text, and simple language for comprehension.
Quad 2	**Sean Hudson:** Sean is a native English speaker. Sean is in the top 25 percent of the class academically and is very strong in science and great at sketching.
	Luis Figeroa: Luis is a native English speaker. Luis is academically strong in science and works well with everyone. He is also a highly empathetic student and a great classroom helper.
	Emily Middleton: Emily is a native English speaker. She gets along with everyone in class and has been seen trying to communicate with Amihan, helping her with her locker, and showing her to her other classes. She understands science well and asks a lot of good questions. Emily is always trying to understand the lesson.
	Amihan Villanueva: Amihan is a multilingual learner. Her overall language acquisition proficiency is at level 1 in all domains. Amihan needs a strong support group in her quad because she needs visuals, short text, and simple language for comprehension.

This was your thought process as you created the first quad.

- Bremen and Lindsey will be good face partners because they get along well in class, which is a key component for creating effective quads. They will be able to support each other on assigned work since they are both academically strong. You will be able to maintain the rigor for these two students.
- Linh Kha and Jean Pierre are a good fit as face partners because Jean Pierre can explain the content to Linh Kha in ways she will understand, determined by Jean Pierre's WIDA scores for speaking.
- Bremen and Jean Pierre are a good match as shoulder partners because, though Jean Pierre is not as strong as Bremen in science, he will still be able to understand the material, interact with Bremen, contribute to whole-quad conversations, and articulate his questions and needs on assignments.
- Lindsey and Linh Kha are a good match as shoulder partners because Lindsey's good content knowledge will enable her to help Linh Kha with her questions. Also, Lindsey is very patient and enjoys looking up Vietnamese words on a translation app to communicate with Linh Kha.

This was your thought process as you created the second quad.

- Sean and Luis will be good face partners because they get along well in class, which is a key component for creating effective quads. They will be able to support each other on assigned work since they are both academically strong. You will be able to maintain the rigor for these two students.
- Emily and Amihan are a good fit as face partners because Emily is patient and will take the time to explain the content to Amihan in ways she will understand using drawings and gestures. You have noticed this interaction between these two students during other class activities.
- Sean and Emily are a good match as shoulder partners because Sean likes to explain concepts and Emily likes to ask questions. Emily likes science and tries to solve things before she asks questions, so Sean is a great shoulder partner for when Emily gets stuck.
- Luis and Amihan are a good match as shoulder partners because Luis also has a good content knowledge and will not mind taking the time to help Amihan and guide her through the activities. He doesn't get upset if he must catch up on work when it appears everyone is moving faster than him in completing the assignments.

Now that you've assigned roles in the two quads, it's time for the next step—creating an organized grouping of students chart.

CREATE AN ORGANIZED GROUPING OF STUDENTS CHART

The purpose of an organized grouping of students chart is twofold: (1) it allows the teacher to seamlessly manage student groups and (2) easily modify them to align with activities and lessons. Figure 4.6 (page 96) contains a sample chart based on a classroom of sixteen students. Line 1 includes the quad you created for Bremen Robinson, Lindsey Pearson, Jean

Pierre Léger, and Linh Kha. Line 2 includes the quad you created for Sean Hudson, Luis Figeroa, Emily Middleton, and Amihan Villanueva. The other lines are for the remainder of your native English-speaking students.

	Organized Grouping of Students: Two Pairs in a Quad			
	A1	A2	B1	B2
1	BR	LP	JP	LK
2	SH	LF	EM	AV
3	SR	LR	VS	JS
4	SG	AV	LE	KT

FIGURE 4.6: Example of a completed chart with fictitious students' initials.

Visit *go.SolutionTree.com/EL* for a free reproducible version of this figure.

Notice that you can create quads using both rows and columns. For example, vertical groups A1, A2, B1, and B2 can form a quad, or horizontal groups 1, 2, 3, and 4 can form a quad based on the assignment. The teacher can easily instruct students to form groups horizontally or vertically. For example, "All students in rows one through four get into your respective quads" or "All students in columns A1 to B2 get into a quad." The chart offers flexibility in this way.

The organized grouping of students consists of columns and rows. The column headings contain A1, A2, B1, and B2 labels (adjusted for the number of students in your class) and the row headings contain a list of numbers. The organized grouping of students example in figure 4.6 is made for a class of sixteen students (four columns by four rows). The reason for using letters like A1 to B2 and not using the letters H, MH, ML, and L is to avoid revealing to the class who is considered high, medium high, medium low, or low.

Figure 4.7 shows how the student labels (H, MH, ML, L) align with the organized grouping of students labels.

The organized grouping of students chart is powerful because it allows for the seamless creation of homogeneous and heterogeneous groups. But its power is in the planning. You must carefully plan student placements using the following guidelines.

- The A1 student must *always* be in the top 25 percent academically in either mathematics or science since this is a STEM class.

- The students labeled A2, B1, or B2 can be selected based on the following factors: academic strength or weakness in mathematics, science, or reading; ability to speak the multilingual learner's language; willingness to be patient or work well with others; or other criteria you feel will create a strong and compatible team.

- The organized grouping of students chart allows you to seamlessly manipulate the groups. For instance, you may have an activity where you want the groups to be homogeneous. In that case, you will tell the class you want all letter column groups of the organized grouping of students chart to sit together. Or, if you want heterogeneous groups, you will tell the class that you want all number row groups of the organized grouping of students chart to sit together. This system of planning lessons and activities requires carefully selecting student groups while remembering to not always place multilingual learners as your low students simply due to language proficiency. To ensure this does not occur, take the time to review your multilingual learners' student language assessment reports so you can provide students with the greatest opportunity to reach their potential. Always remember that language ability does not reflect cognitive ability.

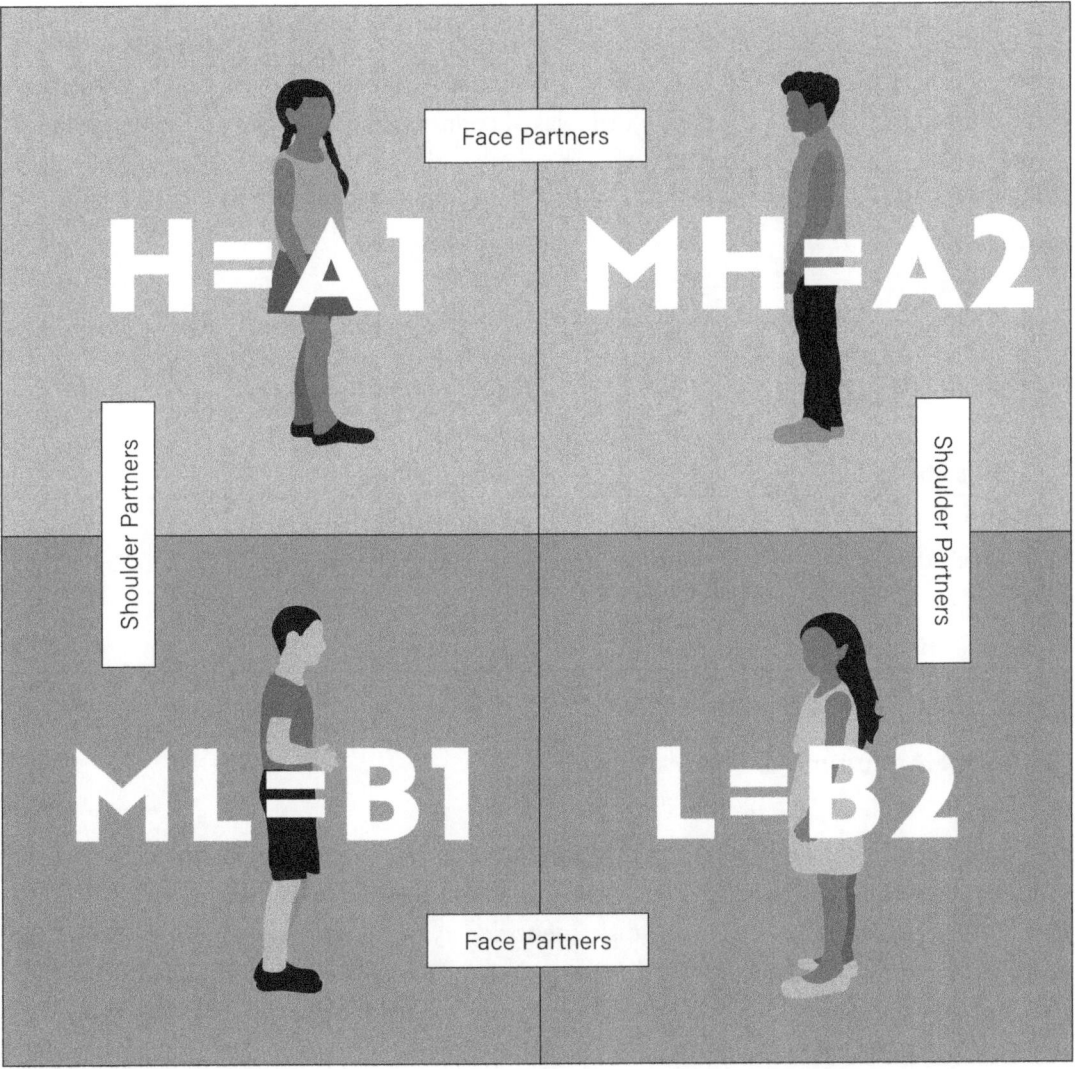

FIGURE 4.7: Aligning the organized grouping of students labels to the quad using shoulder and face partner structure for cooperative learning.

It's important to know that you're not locked into these groups; if dynamics don't work for some reason, you can make changes at any time. Be sure to post the organized grouping of students chart in a prominent location in the classroom for the students to follow.

Imagine the following scenario with your multilingual learners. You just completed the direct instruction component of your lesson explaining the phases of the water cycle. You limited this portion to twenty to twenty-five minutes, based on what you learned about students' attention spans (chapter 2, page 35). The lesson included visuals (printouts and drawings of the water cycle with and without labels of the phases) and time-lapse video of water evaporating and cloud movement. You provided each quad with a list of questions (orally and written) and gave students the opportunity to turn and talk first with their face partners, then with their shoulder partners, then as a quad to answer the questions.

You used Cummins' quadrant to guide the type of activities and questions for this lesson. Cummins' quadrant is the work of researcher Jim Cummins, who created the four-quadrant model to illustrate how the variables of cognitive demand and context impact language learning and how each quadrant progresses from basic interpersonal communication skills (BICS) to cognitive academic language proficiency. Figure 4.8 illustrates Cummins' quadrants.

Quadrant 1	Quadrant 2
Cognitively undemanding	Cognitively undemanding
plus	*plus*
Context embedded	Context reduced
Quadrant 3	Quadrant 4
Cognitively demanding	Cognitively demanding
plus	*plus*
Context embedded	Context reduced

Source: Adapted from Cummins, 1984.

FIGURE 4.8: Cummins' quadrant.

Cummins' quadrant is used to aid in understanding what makes language easier or more difficult for multilingual learners. Difficulty is based on the relationship between the two factors: the cognitive demand of the task, identified as either undemanding or demanding, and the amount of available contextual support, identified as either embedded or reduced (Reiss, 2004). Figure 4.9 contains examples of what students can do in each quadrant (Prokopchuk, 2022).

Next it's time to check for understanding.

You present the students with specific questions to understand their prior knowledge and assess what they gained from direct instruction. You give them silent think time before inviting them to respond. The amount of silent time you should provide students can vary depending on the complexity of the question, a student's age and language proficiency, and the context of the learning environment. Generally, it's recommended to give students at least five to fifteen seconds of silent think time to formulate a response to a question for which they should know the answer (McCarthy, 2018). This brief period allows students to process the question, organize their thoughts, and formulate their responses before sharing with the class. However, for more complex questions or for students who may need additional processing time, consider providing up to two minutes of silent think time to make sense of questions that require analysis to synthesize concepts into a different construct or frame. You can aid this by encouraging journaling, silent reflection, or partner discussions. Giving such chunks of time honors the work being asked of students. After the allotted time, any student can be called on to share their response. Ultimately, it's essential to be attentive to your students' individual needs and adjust the amount of silent time accordingly to support their engagement and learning (McCarthy, 2018).

Quadrant 1	Quadrant 2
Cognitively undemanding plus context embedded. Beginner language levels (1 and 2) students can: • Answer yes-or-no and short-answer questions • Use numbers, symbols, or measurement terms in mathematics • Describe everyday routines or events • Write short notes or create lists • Read signs, symbols, announcements, charts, brochures, simple maps all with the help of visuals, such as illustrations or photos • Understand storytelling with props, gestures, dramatic interpretations	Cognitively undemanding plus context reduced. Intermediate language levels (high 2 to 5) students can: • Follow instructions to conduct simple lab experiments • Complete mathematics calculations, formulas, and questions containing common mathematics terms • Understand or give instructions for specific needs in familiar situations • Understand abbreviations' meanings • Locate basic information in books or electronically • Make personal entries in journals or diaries
Quadrant 3	**Quadrant 4**
Cognitively demanding plus context embedded. Intermediate language levels (3 to 5) students can: • Record procedures from an experiment • Solve problems using manipulatives, charts, graphs, and technology • Represent knowledge in several ways • Analyze information using comparative charts or other organizational tools • Comprehend video presentations, reports, and lectures (with visual clues)	Cognitively demanding plus context reduced. Advanced to fluent language levels (high 3 to 5) students can: • Understand text-dense lectures and most genres and lengthy texts • Read or write reports • Conduct in-depth research • Explain subject-specific concepts from science • Express point of view using subject-specific concepts and terms • Synthesize information from various sources to create presentations • Respond to text-dense multiple-choice questions or open-ended essay questions

FIGURE 4.9: Cummins' quadrant explanation.

By consciously incorporating silent think time into instructional practices, you create a supportive learning environment that empowers your multilingual learners to engage meaningfully with their face partner and shoulder partner. Figure 4.10 (page 100) shows the questions you scaffold to accommodate the language proficiency levels of your multilingual learners without lessening the question's rigor or quality. You provide the questions to each student in the quad and post them on chart paper for the entire quad to write their best answer.

Question	Scaffolded Idea
1. In your own words, what is the water cycle?	1. Offer verbal prompts, cues, and sentence frames to encourage multilingual learners to describe each stage of the water cycle using basic vocabulary and short phrases.
2. Describe the various stages of the water cycle in your own words.	2. Have a visual of the water cycle with areas for multilingual learners to write in the answer and have them identify, list, and number the order and stages of the water cycle and draw a picture of what happens during that stage.
3. Do you think that the water cycle impacts the environment and water quality?	3. Multilingual learners can answer with a yes-or-no response. Provide multilingual learners with sentence starters or frames to help formulate their responses and also provide a picture vocabulary sheet.

FIGURE 4.10: Scaffolded ideas.

These questions are designed to help multilingual learners engage with the topic of the water cycle and develop their language skills while learning about this important scientific concept.

After silent think time, you instruct students to discuss the questions with their face partner. You walk around the room listening to each quad's conversation between face partners. You hear the following interaction from quad 1.

> *Jean Pierre: Yes. I think the water cycle has something to do with the environment and water quality. What do you think, Linh?*
>
> *Linh: Yes. Me too. [Linh Kha then shows Jean Pierre the water cycle visual where she colored the ocean water brown. She then shows him how the brown water from the ocean evaporates to the clouds, the cloud moves, and then rain deposits the dirty water.] Dirty water. [She points to it on the visual.] Goes to clouds. [She traces with her fingers to show movement from the water to the clouds.] Clouds move with dirty water. [She uses her finger again to show clouds moving.] Rain drops dirty water back to Earth. [She uses her finger to trace from clouds to the ground.]*
>
> *Jean Pierre: Do you remember the word for rain going from the ground to the clouds? [While saying this, he traces with his finger from the ground to the clouds.]*
>
> *Linh: [Looks at the other visual aid of the water cycle and points to the word evaporation.]*

Jean Pierre: Yes. That word is evaporation.

Linh: Evaporation.

Jean Pierre: Great job. You said it.

Teacher: How was the water quality in Haiti, Jean Pierre?

Jean Pierre: I did not grow up with problems with clean water, but my father and grandparents told stories about how they had to take buckets to the well to get clean water. They spoke about the long lines and how much work it was to get clean water.

Teacher: Linh, have your parents ever talked to you about the water in Vietnam? [Since Linh scored a level 4 for listening, she can understand the question.]

Linh: Yes. [She draws a picture of a house with rain falling. On the side of the house, Linh draws pipes and large barrels at the end of the pipes with water in them. She points to the buckets.] My grandparents' house. They get water here.

Teacher: Jean Pierre and Linh, thank you for sharing. Linh, your grandparents used a system called rainwater harvesting. I would like to show your picture to the class when everyone is done talking and explain what it is. Would you like to hold up the picture while I talk about it?

Linh: [Nods her head yes and smiles.]

You stop the class to discuss Linh Kha's drawing because it works perfectly with the upcoming STEM challenge. Then you say to the class, "Do you know the source of your drinking water? Discuss with your quad."

After a short discussion, you explain that people today in some communities in continents like Asia and Africa, collect precipitation as a source of fresh water for drinking and daily water use. "The collection of water is called rainwater harvesting. Linh Kha told me that her grandparents use a system like this." You show the students Linh Kha's drawing of her grandparents' home as well as some other pictures of houses with rainwater harvesting systems. This is a great example of a culturally responsive action. It is an opportunity for Linh Kha's culture to be discussed and it also creates an environment for her to feel like a contributor to the discussion.

Then, you turn your attention to quad 2. Prior to creating the quad, you spoke with each of the three native speakers who were selected to be in the quad with Amihan so that they would understand how she would be able to interact with them through speaking, listening, reading, and writing.

Amihan's face partner turns to Amihan, nodding in the affirmative. "Yes. I think the water cycle will carry the polluted water to other places. What do you think?"

Amihan responds, "Me too." She nods her head in the affirmative. She shows the water cycle visual aid provided to her and draws squares to represent pollution and places the squares in every phase of the cycle showing that it gets carried by the water cycle.

Listening to the short conversation between Amihan and her face partner, you rephrase what Amihan drew and ask her yes-or-no questions. "So, Amihan. Does the square represent something dirty, like pollution?"

Amihan nods affirmatively and says yes.

"Is the reason you draw it at every stage of the water cycle because you are trying to show that the water cycle takes the pollution with it?"

Again, Amihan nods in the affirmative.

As you point to the following sentence frames, you ask Amihan which sentence frame she would like to use to express what she means.

1. *The water cycle begins when . . .*
2. *The water cycle is divided into phases; the first one is called . . .*
3. *One important phase of the water cycle is . . .*
4. *The water cycle is a process that helps explain how water moves through the Earth. One way it connects to water pollution is . . .*
5. *Water pollution is when harmful substances get into our water sources, like rivers and lakes. When the water cycle interacts with water pollution . . .*
6. *The water cycle and water pollution are connected. One important concept to understand is that . . .*

These sample sentence starters were created using Magic School (https://app.magicschool.ai/tools), a free artificial intelligence (AI) platform that helps educators create lesson plans, differentiate instruction, write assessments, and much more. Magic School saves teachers time on repetitive, tedious, behind-the-scenes tasks, but for the purposes of teaching multilingual learners, it provides teachers with sentence frames for any topic, translates text to any language instantly, and adjusts text to specific reading-ability levels.

While you are talking to Amihan, you ask her face partner to share his thoughts with his shoulder partner and for Amihan's shoulder partner to interact with Amihan as she selects a sentence frame. By having Amihan select a sentence frame so she can respond and having her face partner and shoulder partner interact with her, they are fulfilling all aspects of the PIES principles of cooperative learning (page 80). To complete this discussion, the quad had to work together to record their final answer to the questions on the poster paper provided to them earlier. This requires collaboration and the ability for the quad to come to a collective agreement for their answers.

The STEM Challenge

Now that you have evaluated student understanding of the water cycle and clarified student misconceptions on the water cycle's connection or lack of connection to pollution, you provide students with a STEM challenge about filtering polluted water. This STEM challenge was amended from Teach Engineering's hands-on activity "Designing Ways to Get and Clean Water" (Shah, Zarske, & Carlson, 2006). As you go through the STEM challenge, notice how the content is scaffolded for Linh and Amihan.

To begin the challenge, each quad selects one country from a list you provide. You provide Linh Kha and Amihan Villanueva translated articles about countries where the people must deal with issues of poverty, water scarcity, and access to clean water daily since the class is assigned a short task to discuss what part the water cycle plays in this issue of polluted water. Including articles and discussions on clean water issues in the world is that it provides relevance. Providing relevance in tasks is twofold: (1) studies show that students who received a more relevant assignment experience more autonomous forms of motivation relative to the students who received a generic or traditional exercise (Johansen, Eliassen, & Jeno, 2023), and (2) students' motivation to complete the task or skill is enhanced because they see meaning in what is being assigned.

Just as the aviation STEM challenge (chapter 3, page 58) was broken into chunks, the same process of chunking the information is used for this water cycle STEM challenge. This challenge comprises five parts.

1. Background knowledge
2. Introduction to the STEM challenge
3. Engineering design phase
4. STEM challenge scenario
5. Evaluation of the STEM challenge

Part 1: Background Knowledge

Part 1 is focused on understanding multilingual learners' background knowledge. Recall that this step is crucial for effective instruction and engagement. This awareness allows educators to tailor their teaching strategies to meet the specific needs and experiences of their students, making lessons more relevant and accessible. When teachers connect new content to what students already know, it enhances comprehension and retention, providing a foundation for more complex concepts. Additionally, acknowledging and integrating students' cultural and linguistic backgrounds fosters an inclusive classroom environment, boosting confidence and participation. Ultimately, leveraging multilingual learners' prior knowledge bridges gaps in understanding and supports their academic success.

- Each team describes the climate of their selected country (tropical, coastal, alpine, desert, and so on). Ask them to brainstorm for two minutes and write down different water sources for their climate.
- Disaster strikes! After two minutes, go around to each group and tell them that they can no longer get water from a major water source they listed. Ask them to brainstorm what they could do if this water source was removed. Provide this update in writing as well and give it to each quad.

This is where the power of the quad comes into play. Linh and Amihan will need to rely on their face and shoulder partners to help them understand what is happening and what they are required to do. If the quad struggles to assist them, the teacher can intervene. You scaffold the activity for Linh and Amihan by providing the following vocabulary definitions for words that you feel will give them difficulty as well as questions to check their understanding.

- Vocabulary:
  - *Disaster*—A sudden event or accident that causes great damage or loss
  - *Major*—Significant or important
  - *Source*—A place, person, or thing from which something originates or is obtained
  - *Brainstorm*—To generate a large number of ideas or solutions to a problem
- Questions:
  - What happens after two minutes?
  - What does the word *disaster* mean?
  - If the water source listed in the text was removed, what would you need to do?

Although you may not be able to provide translations throughout the STEM challenge, you translate the vocabulary words and definitions into Linh's and Amihan's respective languages using Magic School.

Part 2: Introduction to the STEM Challenge

Here you provide students with an introduction to the STEM challenge. As one example, here's how Teach Engineering contributors Jay Shah, Malinda Schaefer Zarske, and Denise W. Carlson (2006) open the activity:

> What would happen if you woke up tomorrow, turned the water tap on for a drink, and dirty brown water came out? How about if you turned the water faucet and no water came out? What would happen if, at home this evening, people came by your house and said that you should not drink your water because a terrible contaminant (pollution) had been found in it? What if people said you had to limit your water use because we were running out? What would you do? This all sounds really bad, but it actually happens!
>
> All over the world, problems of water quality and quantity are very real. In other parts of the world people must walk very long distances every day to get water because it is not piped to where they live. They do not have water faucets or wells. Sometimes their trek to get water is tiring or dangerous. In the U.S., after Hurricane Katrina, water supplies in parts of New Orleans were too dirty for drinking. The water was filled with dirt and sand. You don't want to drink water filled with dirt, do you? Also, regions in the southwestern U.S. have been having a water shortage for years. There just is not enough water for everyone who needs it.
>
> Hundreds of years ago, these problems existed too. What ideas do you think people had a long time ago to fix these sorts of problems? Well, ancient Jordanians developed a highly extensive system of cisterns (underground water tanks) and pipes to collect and distribute water. In Cambodia and other places, people developed ways to naturally clean water on the bottom of riverbeds. Some of these ancient technologies still exist today. Many places in Europe, including Germany, still partly clean water by filtering it through riverbanks.

What do we do when these water problems occur? Environmental engineers, water resource engineers, chemical engineers and many other people work on ways to solve these problems. Engineering teams and companies often develop new technologies to clean up dirty or polluted water and collect and conserve our water sources. Today we are going to act as engineering companies, looking at different situations in which water sources are threatened. We will even design ways to handle specific water problems. Are you ready?

You know that Linh and Amihan need a scaffolded version of this introduction text. Figure 4.11 shows the scaffolded grade 2 reading level you provided for Linh. Linh will read this over with Jean Pierre's assistance as needed to ensure she understands the text. If Jean Pierre needs assistance in explaining it, then Linh will turn to her shoulder partner, Lindsey, for further assistance. Should she need more support, Linh will ask the entire quad. The teacher is the last person to approach.

Water Troubles

What if tomorrow, when you want a drink, the water doesn't look clean? Or what if there's no water at all? What if people said the water at your home isn't safe to drink? What if they also said we don't have a lot of water left? What should you do in these situations? These things may seem scary, but they can really happen!

Introduction

Water is a vital resource that we need every day. But what would happen if we woke up one day and couldn't get clean water from our taps? Or if there was no water at all? These are real problems that people face all over the world. Let's explore some of these water troubles and learn how engineers are working to solve them.

Water Scarcity

In some parts of the world, people don't have access to clean water like we do. They must walk long distances every day just to get water. Can you imagine having to do that? These people don't have water faucets or wells, so they must find water sources far away from their homes. This can be tiring and even dangerous.

Water Pollution

Water pollution is another big problem. After Hurricane Katrina, parts of New Orleans had water that was filled with dirt and sand. It was not safe to drink. Imagine if the water coming out of our taps looked like that! We wouldn't want to drink it. In other places, like some regions in the southwestern United States, there is a water shortage. There simply isn't enough water for everyone who needs it.

Modern Solutions

Today, engineers and scientists are working hard to find solutions to our water problems. Environmental engineers, water resource engineers, and chemical engineers are just a few of the professionals who are dedicated to this work. They come up with new technologies to clean up dirty water, protect our water sources, and conserve water. These engineers and their teams are like superheroes, helping us ensure that we have enough clean water for everyone!

Let's Be Engineers!

Now it's our turn to be engineers! We are going to explore different situations where water sources are threatened and find ways to solve these problems. Get ready to use your creative-thinking skills to design solutions that will make a big difference in the world of water.

Remember, water is a precious resource that we should never take for granted. By learning about water problems and exploring solutions, we can all become water heroes.

FIGURE 4.11: Scaffolded introduction to the STEM challenge, grade 2 reading level.

Figure 4.12 shows the scaffolded grade 1 reading level you provided for Amihan. Amihan, like Linh, will read this over with Emily's assistance as needed to ensure she understands the text. If Emily needs support in explaining it, Amihan will turn to her shoulder partner, Luis, for further assistance. Should she need more help, Amihan will present it to the entire quad. The teacher is the last person to approach.

Water Troubles

What if the water from your tap wasn't clean? Or what if there was no water at all? Some people even have to walk far for water every day. In some places, like after a big storm, the water might not be safe to drink. And in some areas, there's not enough water for everyone. But a long time ago, people had ideas to fix these problems. They used special tanks and pipes to get water or found ways to clean it naturally. Nowadays, there are engineers who work on making the water clean and conserving it. So, we can also think of ways to solve these water problems!

FIGURE 4.12: Scaffolded introduction to the STEM challenge, grade 1 reading level.

Part 3: Engineering Design Phase

In part 3 of this challenge, you review the engineering design process (chapter 3, page 65). You use visual representation of the engineering design process to provide scaffolds for Linh and Amihan. You model the process with a simple task while specifying which part of the engineering design process you are targeting. For example, say you model this as part of a STEM challenge about designing a paper tower. You would take students through the following seven-step process.

1. **Identify the problem:** Create the tallest tower possible using only paper (the paper can be cardstock paper or notebook paper) and tape with a time limit of fifteen minutes. *You explain this by showing the materials and using other visuals to assist like a video or pictures of students doing the STEM challenge.*

2. **Research or gather information:** This step requires that they look up similar designs and learn about basic engineering concepts like strong bases and structural integrity. *Model this by having a few books on the topic or show them how to research it online.*

3. **Brainstorm ideas:** Think of different structures and shapes (for example, cylinders or pyramids) that could be used to maximize height while ensuring stability. *Create a variety of towers of different shapes and ask simple questions to get feedback, such as, "Which looks stronger?" and "Which do you think can go higher and stay stable and not fall over?"*

4. **Create a plan:** Sketch the final design with dimensions, the arrangement of paper rolls or folded sheets, and key features that contribute to the towel's strength. *Make sketches of your ideas to demonstrate that this is what they will also need to do (for example, including measurements on the diagrams).*

5. **Create a prototype:** Build a small-scale version using paper and tape to see how well the structure holds up and identify potential flaws. *Create the prototype according to your sketch. Point to or state the flaws you notice after creating the prototype.*

6. **Test prototype:** Measure the prototype's height and observe if it remains standing under different conditions like a gentle breeze or slight pressure. *Blow the prototype with a fan or shake the table to test if it can withstand those forces. Write notes for the students to model noting the outcomes of each iteration.*

7. **Analyze results and modify:** Evaluate the prototype's performance. If the tower collapses or isn't as tall as expected, modify the design by reinforcing weak points or adjusting the structure's shape and stability. *Review notes and make changes to the prototype and go back to step 6. Model this by showing the students that you are going back and trying again.*

You can also use icons to represent each step. For example, for "identify the problem" you can use a question mark icon or for "gather information" you can use a picture of a person reading with a lot of books around them, and so on. These will provide visuals to help them understand the steps and process.

Part 4: STEM Challenge Scenario

You provide each quad with a water problem scenario card for their STEM challenge. The scenario card informs them of their region and the problem that has occurred with the water.

To scaffold for Linh and Amihan, you have the scenarios adjusted for their reading level. You also provide the quads with silent think time, which offers Linh and Amihan time to understand the scenarios and identify any questions they may have.

Figure 4.13 shows a sample scenario.

A. Providing Clean Drinking Water for a Coastal Community	B. Providing Clean Drinking Water for a Coastal Community
This community lives in a region that has clean water sources. However, there are just too many people living in the area for the water sources to provide enough clean drinking water for everyone. The community is in a coastal region by a saltwater ocean. Design a system for the community that could help them provide enough clean drinking water for so many people.	The people in this community live near the ocean and have some clean water, but there are a lot of people in the area. This means they don't have enough clean water for everyone to drink. If you were to help this community, how would you make sure there is enough clean water for all the people?

FIGURE 4.13: Water problem scenario showing the original scenario (A) and the scaffolded scenario (B).

You allow each team time to brainstorm a way to fix the problem. You let them know that any design or method is fine, except they must make it realistic for their situation by considering the resources and limitations of the climate in their scenario. Encourage them to be creative!

You provide the following scaffolds for Linh and Amihan.

- You provide Linh and Amihan with the scenario ahead of time. In order for them to understand how to brainstorm, you model the process using a simple problem. Since Linh's listening skills are at a level 4, she does not need as much scaffolding and modeling as Amihan, since her listening skills score is level 1, but her actual score is 1.7, which is between levels 1 and 2.
  - As needed, guide and evaluate them with questions like: How does the design relate to the climate for this scenario? Why did you choose your design? What sort of power or energy is needed for your design to work (batteries, electricity, or nature)?
  - Ask simple questions and use simple sentences. You provide sentence stems and sentence starters for them to respond.
  - Have teams draw their engineering method(s) to fix their water problem scenarios on large paper using markers or colored pencils.
- Since this step requires drawing, it is not as challenging for Linh and Amihan.
  - Ask students to label the parts of their designs and be as detailed as possible. Point out how designing, planning, and creating a model are steps in the engineering design process.

Part 5: Evaluation of the STEM Challenge

As a means for students to determine whether their designs are just limited to their country and climate or if it can be used as a global solution, students receive the opportunity to compare their solutions with other situations. They may also come to realize that if they make a few changes in their prototype, perhaps their solution can help others in many parts of the world with different conditions. Therefore, they are making connections to how scientists' ideas for one part of the world can impact real-world situations in other parts of the world.

Instruct each group to switch their original family scenario note card with another group.

- Ask students to evaluate their current design for this new scenario.
- Have students discuss if and why their design would or would not work in this different situation in a different climate.
- Ask them to think about what they could do to make it work for the new conditions.

Once again, since Linh has a higher listening level, she does not need as much scaffolding as Amihan for understanding the action needed.

- Both will benefit from using simple sentences to explain what actions are required. You provide a picture of the climate for the new scenario. You point to the picture and ask if what they drew would work if it was in this climate.

- You ensure that your questions only require simple one- or two-word responses, yes-or-no answers, or gestures. For example, you ask them to think about what they can do to make it work in the new climate and to draw what they are thinking.

Another option would be to have them create a prototype. By using a prototype, they can make the changes to demonstrate what adjustments they would make based on the new scenario's climate.

STEM challenges can be scaffolded and amended so that all students can successfully participate. All students have assets to bring to lessons, as seen by how Jean Pierre and Linh provided feedback on their experiences with water where they grew up. Using language proficiency as the gatekeeper to decide who can participate or not is not an acceptable system. If time is taken to scaffold STEM challenges for multilingual learners, then all students will have equal opportunities to participate in a field that has the power to change lives and livelihoods.

Key Takeaways

Creating specific and targeted student groups allows educators to tailor learning experiences in STEM to the unique needs of each student, ensuring inclusivity for all, including English learners. By grouping students strategically, as in using Two Pairs in a Quad, educators can foster a collaborative environment where diverse language proficiency levels are not barriers but, rather, assets. Students with stronger English and academic skills can support their peers in comprehension, while hands-on, visual, and inquiry-based activities transcend language barriers and engage all learners effectively. Such purposeful grouping encourages peer learning, mutual respect, and a supportive classroom culture, making STEM concepts accessible and enriching for everyone, regardless of language proficiency.

- The Two Pairs in a Quad model brings real language into the STEM classroom. It allows multilingual learners to hear language, content, and vocabulary in its real context simultaneously.
- The Two Pairs in a Quad model provides opportunities for language to be used for sensemaking and communicating about science in a small, safe environment.
- Two Pairs in a Quad is one way to create inclusivity in the multilingual classroom. The process requires practice and fine-tuning as you prepare to pair up students in a quad.
- Cooperative learning strategies aid in designing activities that are engaging and are not content specific.
- Cooperative learning strategies can provide the following outcomes for your multilingual classroom.
    - A positive attitude
    - More interaction with peers

- A sense of safety in the classroom for taking risks in their learning and speaking, reducing the need for constant teacher dependence
- Positive emotional sensitivity

• The engineering design process in STEM challenges encourages creative problem solving, critical thinking, and iterative learning, enabling students to approach challenges methodically and develop practical solutions while reinforcing key STEM concepts.

CHAPTER 5

Leveraging Student Assets and Building Content Knowledge Through Scaffolding

> *If students never get past basic skill building, they will never get a chance to flourish at high levels in the areas where they have the most talent. Limiting students in this way is an injustice and a waste of valuable student assets. All students have strengths and deserve opportunities to thrive.*
>
> **—AVID OPEN ACCESS**

Leveraging student assets and building content knowledge through scaffolding, especially for multilingual learners, means creating a classroom environment where every student's unique linguistic, cultural, and experiential backgrounds are viewed as valuable resources. By recognizing and incorporating these diverse strengths into instructional strategies, educators can scaffold learning in a way that connects prior knowledge to new concepts. This approach makes rigorous content more accessible, allowing multilingual learners to confidently engage with complex STEM topics while enhancing their language skills. Ultimately, this inclusive strategy empowers all students to progress academically, fostering an atmosphere of collaboration and shared growth.

So, how does STEM leverage student assets?

- Each student comes with funds of knowledge; STEM projects and tasks provide multiple ways to solve a problem through their experiences.

- STEM creates conditions for students to collaborate and engage in productive discourse. This is an important component of language development for multilingual learners because they get the opportunity to hear the language and vocabulary in their proper context.
- STEM challenges offer multiple correct answers and pathways to the solution, exposing students to multiple ways of thinking and numerous modalities to express themselves.
- STEM offers every student a voice and, therefore, makes them an asset to the STEM task outcomes and solutions.
- STEM provides students opportunities to fulfill their need to learn how to analyze, reason, and make complex decisions to improve their chances of success later in life.

STEM and its independent disciplines allow multilingual learners to engage in substantive conversations about what they are learning, to make connections between spoken and written practices and meaningful artifacts of the discipline, and problematize knowledge and question accepted wisdom (Gibbons, 2007). Remember that language is a product of interaction and learning, not a precursor or prerequisite (Gibbons, 2007). The Next Generation Science Standards (NGSS), adopted by many states in 2015, represent not only a new set of benchmarks against which to measure students' content knowledge, but a completely new three-strand approach called *three-dimensional learning* (*3-D learning*) that offers an opportunity for students to develop skills critical for success in STEM areas.

The elements of 3-D learning are: (1) disciplinary core ideas, the foundational concepts of science; (2) science and engineering concepts, the methods and practices used by scientists and engineers in understanding the world; and (3) crosscutting concepts, which encapsulate overarching themes, such as patterns and systems that are applicable across the various science domains like life sciences, physical sciences, Earth and Space sciences, and the engineering and technology domains. This new three-dimensional approach to teaching and learning identifies the acquisition of scientific practices and the understanding of cross-disciplinary concepts as equally important as learning the actual science content (Teed, 2020).

Several science and engineering concepts, such as developing and using models; constructing explanations; engaging in argumentation based on evidence; and obtaining, evaluating, and communicating information, were all identified as practices that present significant challenges for multilingual learners (Quinn, Lee, & Valdés, 2012). Adding to the challenges facing teachers of multilingual learners, the content within the NGSS is carefully organized and spiraled from kindergarten all the way to twelfth grade (Teed, 2020). This means that multilingual learners who often experience gaps in their education may lack background knowledge necessary to master grade-level standards.

In this chapter, I describe the unique assets multilingual learners bring to the classroom and explain the connection to funds of knowledge. Multilingual learners have the best chance of reaching their potential when teachers intentionally connect their assets to instruction as they build academic vocabulary, tap students' funds of knowledge, and activate and assess prior knowledge. With the help of fictional student Fatou Alaoui, I will share five strategies you can use to practice these principles through scaffolding. Toward the end of the chapter, I will introduce a STEM challenge that illustrates how to use scaffolding to tap students' funds of knowledge and leverage multilingual learners' assets.

Terms Defined

According to Katherine Roe (2019), author of *Supporting Student Assets and Demonstrating Respect for Funds of Knowledge*, *student assets* are a student's culture, community, and family. *Funds of knowledge* are the essential cultural practices, bodies of knowledge, and skills gained from life experiences and developed from within their communities and families (Lopez, 2016; Moll, 2024; Moll, Amanti, Neff, & Gonzalez, 1992; Volman & 't Gilde, 2021). Using a funds-of-knowledge approach in lesson design allows teachers to understand students' overall abilities and experiences, which can help teachers draw on these skills to enrich students understanding of academic content while also motivating them during classroom activities (Washington Office of Superintendent of Public Instruction, 2023). The more teachers learn about students' assets increases the students' sense that they are a part of the classroom community. Leveraging student assets means teaching with an asset-based approach. According to NYU Steinhardt (2018), an *asset-based approach* focuses on students' strengths rather than their weaknesses or deficits. It views diversity in thought, culture, and traits as positive assets. Students are valued for what they bring to the classroom rather than being characterized by what they may need to work on or lack. Asset-based teaching seeks to unlock students' potential by focusing on their talents.

In this chapter, we examine how students' lived experiences are an asset to the academic content they encounter in STEM instruction. The beauty of STEM is that it connects to everyday life and authentic situations that students may have interacted with or experienced. STEM's built-in approach to solving problems makes the learning more intriguing and relevant. But you will still find that many educators, teachers, and administrators are skeptical that young children can learn just by diving into a problem without prior practice on the skill needed (Klein, 2022). Specifically, in relation to multilingual learners, the concern is often their prior knowledge about STEM subjects. When students enter U.S. schools, they are not typically assessed for their content knowledge. Instead, a student's identification and course placement, at least at the secondary level, is typically determined by their level of English proficiency (Francis & Stephens, 2018). The 2018 report by the National Academies Press determines that it is imperative that multilingual learners have the same quality of STEM-related learning opportunities as their native-English-speaking peers because multilingual learners bring a wealth of resources to STEM learning. This includes their knowledge and interest in STEM-related content born out of experiences in their homes and communities, home languages, variation in argumentation practices, and, in some cases, experiences with schooling in other countries. In other words, each student has various life experiences that can be a unique asset to solving the problem.

A great example of the value and importance of funds of knowledge can be seen in research from the 1990s. Researchers at that time sought to investigate how Latinx (a gender-inclusive identifier that recognizes LGBTQ+ individuals with Latin American roots) families living in a sparsely populated area in the southwestern United States–Mexico borderlands with little access to affordable health care gained and shared knowledge to respond to their economic, environmental, and social landscape (PBS LearningMedia, 2022). In this vacuum, families developed remedies and first-aid procedures based on their ancestral folk medical knowledge. This ancestral folk knowledge was this community's funds of knowledge. This wealth of knowledge had a strategic importance for economic well-being and survival for Latino families in that environment. Other examples of funds of knowledge include the following.

- Making gnocchi from scratch
- Keeping score at a curling match
- Quilting
- Spinning wool into yarn
- Fixing a car
- Caring for a crying baby
- Preparing a seder
- Braiding cornrows

Funds of knowledge are powerful because they are culturally relevant to students. They bring more diversity into the classroom for students who don't relate to the dominant culture. It offers teachers a chance to become researchers of their students' lives, which allows for better connections with students' home cultures and ultimately acts as a bridge when needed. And it creates a deeper student connection to learning materials and classroom activities (McDonald, 2018).

Funds of Knowledge and Instruction

Although students come to the classroom with varying funds of knowledge, they also have varying degrees of language proficiency and academic experiences. Assessing and building on funds of knowledge can be a challenge for teachers, especially if a student's education has been disrupted. Although many multilingual learners do come to school with the ability to think, read, and write in their native language, it is not uncommon that their native language proficiency varies due to disrupted schooling, the use of several languages spoken in their home, or limited exposure to literature in their first language (Sousa, 2010).

A teacher's ability to build on funds of knowledge is especially important for multilingual learners who might not realize they have prior knowledge about a subject because the new learning is in English (Fleenor & Beene, 2019). Because multilingual learners present a unique set of challenges to educators due to the central role that academic language proficiency plays in the acquisition and assessment of content-area knowledge, multilingual learners need activities that activate deep thinking and set the stage for the lesson that follows (Francis, 2006).

The benefits of focusing on assets as opposed to focusing on deficits are profound. According to AVID Open Access (n.d.a), when students equate themselves with their deficits and failures, they can be easily defeated. This can lead to disengagement from school, a disinterest in learning, and a loss of self-esteem. On the other hand, when students recognize their talents and are encouraged to use them successfully, they thrive, developing feelings of pride and self-confidence as well as a desire to learn more. Let's take a closer look at these twelve ways to build on student assets (AVID Open Access, n.d.a).

1. **Voice and choice:** Although all students in your class must work on the same standard, students can choose their topic (having a voice in their learning) and decide on how they will practice or apply the new concept (having a choice in demonstrating their understanding).

2. **Cultural relevance:** As discussed in chapter 3 (page 51), it's important for teachers to bring cultural relevance to their practice. Allowing students to draw on their personal and lived experiences and apply those experiences to their learning provides students with enough voice and choice so that they can make personal connections. The more that students are empowered to bring in their personal and cultural experiences, the more that they will connect meaningfully with the learning.

3. **Differentiated pathways:** There are three places where a teacher can differentiate instruction: (1) content (what students are expected to learn), (2) process (activities students use to master the content), or (3) product (the method students use to demonstrate learning; KNILT, 2021). Since not all students learn in the same way, you should embed different learning options throughout the lesson or unit. For product, if a student suffers from test anxiety, maybe they can explain what they know in a conversation. For process, if a student doesn't write well, perhaps they can record an audio response, and for content, offer both a text-based and a video version of the information. The more flexibility that you can provide along the way, the more likely that students will be able to play to their strengths and be successful.

4. **Project-based learning:** This is the pinnacle of voice and choice. Students have greater flexibility to make it their own, connect to personal interests, and show off their talents because of the open-ended nature of project-based learning. This is a nice blend of voice and choice, cultural relevance, and differentiated pathways.

5. **Inquiry learning:** Like project-based learning, inquiry learning allows students voice and choice in their learning. Inquiry learning provides opportunities for students to generate and find answers to their own questions, where they can make connections between the content and the process.

6. **Interdisciplinary learning:** The NGSS framework is an interdisciplinary approach to teaching science. It incorporates life science, Earth and space science, physical science, and engineering. This approach increases the

authenticity and meaningfulness of the learning, offering context and connection among the various academic areas and providing insights into how all learning is connected. When multiple content areas are woven together into one project (STEM, for example), the odds are increased that students will find an area of strength embedded in the learning experience.

7. **Just-in-time enrichment:** Reteaching lessons from students' last grade level is not always a possibility even if the students need to first be retaught those prerequisite standards. But, if you can develop a list of enrichment activities and challenges ahead of time, you will be able to quickly apply them to a lesson at a moment's notice. Be sure that these just-in-time enrichments are flexible enough to allow students to integrate their talents and strengths.

8. **Enrichment stations:** Many teachers have stations for students to visit while the teacher is working in small groups. If you make one of the stations students visit the enrichment station while you are teaching in small groups, every student gets the opportunity to experience an enrichment station.

9. **Playlist options:** Playlists are checklists and progressions of learning tasks. They are customized to your learners and offer voice and choice. Playlists are a sequential list of tasks and learning experiences that each student is expected to complete to achieve a learning objective (AVID Open Access, n.d.b).

10. **Genius hour:** Genius hour is an opportunity for students to pursue personal interests and passions. During this dedicated time, students set their own goals and develop a pathway toward achieving them.

11. **Makerspace:** What would a STEM classroom be without a makerspace? Makerspace is a space where students can make things. It's a playground of making with very few rules or parameters. Makerspaces usually contain various materials and the tools necessary to create something.

12. **Creation:** Creation involves complex, higher-order thinking and is a very authentic means of enrichment. Creation is another form of makerspace, but if students don't get the opportunity to visit a makerspace because it is in some other part of the school, they miss out and don't get the enrichment experience. However, if you can embed creation into core classes, all students will get the chance to create, providing a more equitable approach to student creation and enrichment.

For more on AVID's (n.d.a) twelve ways to build on student assets, visit https://avidopenaccess.org/resource/accelerate-learning-by-building-on-student-assets.

According to the IRIS Center (2022c), tapping into multilingual learners' background knowledge is among the more effective instructional techniques available for activating their funds of knowledge because it helps students make personal connections to new information, which in turn helps them understand the concepts. There are three ways to do this that will make the twelve ways to build on students' assets activities more meaningful for multilingual learners in your class. The three ways are: (1) build academic vocabulary,

(2) activate funds of knowledge, and (3) assess prior knowledge. Including these in the activities will help to highlight student assets by providing them with opportunities to flourish at high levels in the areas where they have the most talent. Let's look at each of these in more detail in the following sections.

Build Academic Vocabulary

One of the greatest debates around teaching multilingual learners is: Which comes first, the vocabulary term or its definition? There are two competing hypotheses.

1. **The skill building hypothesis:** This view asserts that vocabulary definitions should be taught prior to the lesson (Bilash, 2011). Students learn language by first learning grammar rules and memorizing vocabulary. We make these rules of new words automatic by producing them in speech or writing, and we fine-tune our (conscious) knowledge of grammar and vocabulary by getting our errors corrected (Krashen, 2017).

2. **The comprehension hypothesis:** This view argues that students should make sense of the word first before they receive the definition. Students acquire language when they understand what they hear or read; their mastery of the individual components of language (skills) is the result of getting comprehensible input (Krashen, 2017).

I lean on the side of the comprehension hypothesis because learners best acquire language when given appropriate input, but the caveat is to infer meaning *just beyond* their level of language competence, Vygotsky's (1978) zone of proximal development. The zone of proximal development supports this hypothesis where students must go beyond what they already know and build their new understanding on that foundation (Texas Education Agency, 2020). If students receive the opportunity to make sense of the situation first using their own words, thoughts, and ideas, they will have a better understanding of the concept. So, when the vocabulary word is introduced, it will be easier to understand because they have been provided comprehensible input.

Consider the following example—from professor and author Pauline Gibbon's (2014) book, *Scaffolding Language, Scaffolding Learning, Second Edition*—of going just beyond a student's zone of proximal development:

> Imagine a child learning to feed or dress themselves. At first the adult has to perform the whole process. Then the child gradually performs parts of the process, with the parent still assisting with the more difficult parts. Finally, the child is able to do the entire process unaided. This is an example of successful coordination between the parent and the child. This assistance led the child to reach beyond what they were able to achieve alone and will provide the child with the ability to participate in new tasks. (pp. 13–14)

Now think of this in terms of your multilingual learners and how they will learn new ways of using language to understand content and vocabulary. By designing lessons with scaffolds, teaching vocabulary and content, and ensuring that language is used for the purpose of making meaning and communicating, students have opportunities to build on previous knowledge to develop new knowledge. Inserting these opportunities into lessons in the form of external dialogue and student interactions in science lessons aid the development of thinking and language learning (Gibbons, 2015).

In teaching, the more that students understand and can make sense of the material and vocabulary, the more likely they will be to understand the content and context of the lesson (English Learners Success Forum, n.d.). Teachers can support this by providing visuals, as well as leveraging student assets by incorporating cultural backgrounds and funds of knowledge, taking the time to build background knowledge, and always assessing prior knowledge. Using visuals is crucial when teaching academic vocabulary to multilingual learners because it bridges language barriers and provides concrete references for abstract terms. Scan the QR code and watch my video about 3-D word problems where you can see the power of using visuals in action.

Word Problems for English Learners
https://youtu.be/9Nh4ByzRehY

Visual aids such as pictures, diagrams, videos, and graphic organizers enhance comprehension by offering contextual clues and supporting retention through dual coding, where information is processed both visually and verbally. These tools make complex concepts more accessible, reduce cognitive load, and cater to diverse learning styles, ensuring that multilingual learners can grasp and apply new vocabulary effectively. Integrating visuals into instruction also fosters an inclusive classroom environment where all students can participate and succeed, regardless of their language proficiency.

In the following sections, we discuss tapping funds of knowledge and assessing prior knowledge in more detail.

Tap Funds of Knowledge

Students' funds of knowledge can bring a richness to STEM discussions that isn't present in a class of all native English speakers. Tapping students' funds of knowledge also has multiple benefits for academic and language development because:

- STEM creates conditions for students to collaborate, which is an opportunity for the multilingual learner to hear the language and vocabulary in its proper context
- STEM projects have multiple correct answers and pathways to the solution, which exposes students to multiple ways of thinking
- STEM participation reaffirms that every student has a voice and is an asset to the task
- STEM provides students opportunities to learn how to analyze, reason, and make complex decisions to improve their chances of success later in life (Klein, 2022)

Funds of knowledge—which encompass the accumulated cultural, experiential, and practical knowledge of communities and individuals—can be instrumental in enhancing STEM learning for multilingual learners. By integrating these rich repositories of knowledge, educators can contextualize STEM concepts within the familiar lived experiences and cultural narratives of multilingual learners, making abstract principles more tangible and relevant. Using funds of knowledge not only fosters deeper comprehension by anchoring STEM concepts in students' preexisting knowledge frameworks, but also promotes a sense of belonging and confidence, enabling multilingual learners to see themselves as valuable contributors and active participants in STEM fields. Consider the following strategies for tapping into students' funds of knowledge.

- **Learn about students' backgrounds:** Take the time to understand students' cultural, linguistic, and personal experiences through surveys, interviews, and family engagement activities (see chapter 3, page 41).
- **Build relationships:** Foster strong, respectful relationships with students and their families to create a foundation of trust and open communication.
- **Connect curriculum to real-life experiences:** Design lessons that relate to students' everyday lives, interests, and cultural practices. Use examples and scenarios that resonate with their backgrounds.
- **Incorporate diverse perspectives:** Include materials and resources that reflect students' cultural diversity. This might involve using texts, case studies, and examples from various cultures and communities.
- **Encourage student contributions:** Create opportunities for students to share their knowledge and experiences in class discussions, projects, and presentations. This validates their backgrounds and enriches the learning environment for everyone.
- **Collaborate with families and communities:** Engage families and community members as resources and partners in education. Invite them to share their expertise and experiences in the classroom.
- **Differentiate instruction:** Adapt teaching strategies to meet students' diverse needs, leveraging their prior knowledge and experiences to make learning more relevant and accessible.

Activate and Assess Prior Knowledge

Activating and assessing prior knowledge can be challenging for teachers of multilingual students because the students come to classrooms with varying degrees of prior knowledge, academic experiences, and language proficiency.

According to the IRIS Center (2022c), tapping into multilingual learners' backgrounds is among the more effective instructional techniques available for activating prior knowledge because it helps students make connections to new information, which in turn helps them understand the concepts. When teachers make connections between the students' background and the lesson, they validate the students' culture and experiences. Making this connection can also help facilitate a greater interest in the lesson. For example, in a class unit on severe weather, a student may not know what causes a tsunami, but they may have experienced one or know of a person or family member who has. These funds of knowledge can add to the discussion and make the multilingual learner a contributing member of the classroom. Therefore, it is important to know a students' background so you can use what they already know to enrich the STEM topic or challenge.

Assessing students' prior knowledge enables you to tailor your teaching to their specific needs. These straightforward and flexible strategies can quickly provide insight into what your students already know and where they may need additional support. Consider the following six strategies that can be used interchangeably for activating and assessing prior knowledge.

1. **Anticipatory guides:** Create anticipation guides with statements related to the upcoming topic. Students indicate whether they agree or disagree with each statement, which activates their existing knowledge and sets a purpose for learning.

2. **Think-pair-share:** Ask students to think about what they know about a topic, pair up with a partner to discuss their thoughts, and then share their ideas with the class. This encourages collaborative learning and activates prior knowledge. This is best done within their quad so the multilingual learner has the supports from their face partner and their shoulder partner.

3. **Graphic organizers:** Use graphic organizers like concept maps or mind maps where students can visually display their prior knowledge about a topic. This helps them organize their thoughts and make connections. The sketch of the airplane in chapter 3 (page 72) is a perfect example. The students labeled the parts of the plane on their diagram. The students working in their quad were able to brainstorm with each other about the parts of the plane while labeling them, and the multilingual learner was also allowed to label the parts in their native language. This is also a learning experience for the native English speakers to learn those terms in another language.

4. **Brainstorming sessions:** Conduct brainstorming sessions where students share what they know about a topic. Write their ideas on the board to create a collective knowledge base. Let's use a kindergarten geometry lesson as an example. The teacher might print out four large pictures of a playground on a laminated poster-sized paper for each quad. Ask the students, using a marker, to identify all the shapes they recognize on the poster paper and then to find an object in the classroom with the same shape. Let students know they should be ready to name the shape as a circle, square, triangle, rectangle, hexagon, cube, cone, cylinder, or sphere.

5. **K-W-L charts:** Have students complete a K-W-L chart where they list what they already *know*, what they *want* to know, and (later) what they have *learned* about a topic. This helps students connect new information to their existing knowledge. You can also add *how* they can learn this to create a K-W-H-L organizer, which shifts ownership of learning to students.

6. **Entry tickets:** Most teachers know about exit tickets, but not many know or use entry tickets. At the beginning of class, instruct students to answer a question related to the day's lesson. This provides a quick snapshot of their prior knowledge. For the multilingual learner, you can have the question written in their home language or use visuals if applicable.

Meet Fatou
https://youtu.be/cf3iVO4QwgM
?feature=shared

Meet Fatou

You have a new multilingual learner in your class, Fatou Alaoui. See Fatou's bio and academic profile in figure 5.1 (page 122). Scan the QR code to view a message from Fatou.

Fatou Alaoui

Fatou Alaoui was born in Juba, the largest urban city in South Sudan. Fatou is thirteen years old, but her education is disrupted due to the following factors (Borgen Project, n.d.).

South Sudanese women and girls are less likely to complete primary and secondary education than boys. According to the World Bank, it is estimated that seven girls per ten boys attend primary school. Meanwhile, only five girls per ten boys enroll in secondary education. Although some girls do manage to make it to secondary school, not many of them are able to finish. In 2013, only 500 girls in the entire country were in their graduating year of secondary school.

While education is technically free for South Sudanese students, there are many expenses that the system does not cover. Families are expected to pay additional fees if they want their children to have an education. This includes charges for textbooks, uniforms, school fees, and more. Thus, socioeconomic status plays a major factor in access to education. Although English is the official language of South Sudan, Fatou speaks mainly Juba Arabic, a dialect far removed from standardized Arabic and named for the South Sudanese capital.

Name	Country of Birth	Primary Language	Level of Language Proficiency	Age	Grade
Fatou Alaoui	South Sudan	Juba Arabic	Level 3: Developing	Thirteen	Fifth
WIDA Results					
Listening: 2.9		Speaking: 2.2	Reading: 1.6		Writing: 1.9
Student Capabilities By Level					
Listening (proficiency level 3): Students can understand messages or directions involving language related to routines and familiar experiences. They can better understand spoken language when supported by visual aids, such as pictures, diagrams, charts, and gestures, which help clarify the meaning of more complex or unfamiliar content. Students can identify specific information from audio sources, such as announcements, instructions, and educational videos, particularly when the content is related to familiar topics and presented in a structured format.					
Speaking (proficiency level 2): Students can communicate ideas using words and phrases related to everyday routines or situations. They can retell stories or content-related events; state procedural steps; give reasons why or how something works using diagrams, charts, or images; and state opinions based on experiences.					
Reading (proficiency level 2): Students can understand written texts that include visuals and may contain a few words or phrases in English. They can express ideas or concepts using text and illustrations and share personal experiences through drawings and words. Students at this stage can label steps in processes presented in graphs or short texts and state opinions or preferences through text and illustrations.					
Writing (proficiency level 2): Students can communicate in writing using visuals and symbols that may contain few words in English. They can list procedural steps across content areas, use key terms related to phenomena, and state reasons for particular points of view.					

FIGURE 5.1: Fatou Alaoui's bio and academic profile.

Fatou's English language proficiency level is at the speech emergence stage—level 3 of the five levels of language proficiency which are preproduction, early production, speech emergence, intermediate fluency, and advanced fluency (Krashen & Terrell, 1998). Students at the speech emergence level have good comprehension, can produce simple sentences, and will make grammar and pronunciation errors.

Scaffolds in Practice

What does it look like to plan STEM instruction, keeping in mind what you've learned about funds of knowledge in this chapter? The following are important considerations when designing or selecting a STEM challenge.

- **Begin with a warm-up activity:** All lessons should begin with a warm-up that will assess prior knowledge for the lesson, is aligned to the unit, and get students engaged and ready to learn.

- **Use targeted language:** Ensure that language is used for the purpose of sensemaking and communicating about science where students build on previous knowledge to develop new knowledge.
- **Utilize multiple registers and modalities:** A variety of registers (whole group, small group, pair work, and individual thinking time) and modalities (gestures, pictures, symbols, graphs, tables, equations, oral language, and written language) help multilingual learners communicate ideas with peers and teachers because they benefit from multimodal interactions and diverse opportunities to learn (Francis & Stephens, 2018). Recognizing the importance of multimodality in STEM content areas reorients the focus from what multilingual learners' lack in terms of language to the diverse meaning-making resources they bring to the classroom (McVee, Silvestri, Shanahan, & English, 2017).
- **Preplan strategies:** Preplanned strategies in teaching are foundational for creating an effective, responsive, and dynamic learning environment. They serve as a guide, ensuring that teaching remains focused, purposeful, and aligned with the desired educational outcomes. Plan the strategies in advance and place them in your lesson plan.

Select Strategies to Accelerate the Learning Process

There are five research-based principles of second language instruction that form the basis for multilingual learners' education in grades preK–12. Teaching by these principles is helpful for *all* students but is essential for multilingual learners and struggling learners (Levine, Lukens, & Smallwood, 2013).

The following are the five principles of second language instruction.

1. **Focus on academic language, literacy, and vocabulary:** Teach the language and language skills required for content learning.
2. **Link background knowledge and culture to learning:** Explicitly plan and incorporate ways to engage students in thinking about and drawing from their life experiences and prior knowledge.
3. **Increase comprehensible input and language output:** Make meaning clear through visuals, demonstrations, and other means, and give students multiple opportunities to produce language.
4. **Promote classroom interaction:** Engage students in using English to accomplish academic tasks.
5. **Stimulate higher-order thinking and the use of learning strategies:** Explicitly teach thinking skills and learning strategies to develop effective, independent learners.

Educators employ these five core principles along with targeted teaching and learning strategies to expedite the learning process, enhance academic performance, and cultivate academic language acquisition.

Incorporate Scaffolding Into Strategies

Scaffolding, first used by psychologist Jerome Bruner in the 1960s, sets students up for success by not allowing them to fall (Structural Learning, 2021). *Instructional scaffolding* is a process through which a teacher adds supports for students to enhance learning and aid in the mastery of tasks by systematically building on students' experiences and knowledge as they learn new skills (IRIS Center, 2022b). It is one of the tools that teachers of multilingual learners have at their disposal to move learning forward (Fisher, Frey, & Almarode, 2023).

Researchers find that encouraging students to challenge themselves in grasping new concepts within their zone of proximal development leads to success in learning (Staake, 2021; Wood, Bruner, & Ross, 2006). Learning will take place only when a teacher's support is needed because then that student is working within their zone of proximal development (Gibbons, 2015). When we consider *effect size*—a statistical concept that measures the strength of the relationship between two variables on a numeric scale—scaffolding has been found to have an effect size value of 0.82, which demonstrates that it has been proven to be highly effective (Visible Learning, 2018). However, according to Fisher and colleagues (2023), to move from research to reality and actualize the potential for scaffolding to accelerate student learning, we must implement scaffolding in a way that offers the right level of challenge for students because there is the risk of underscaffolding, overscaffolding, and failing to remove scaffolds once learning has occurred.

Here are five easy and effective strategies you can use to assess and build your students' funds of knowledge regarding the content while checking for and building vocabulary that align with the students' level of language proficiency.

1. Picture-word match
2. The slow reveal
3. Flexible vocabulary journal
4. Pose, Pause, Pounce, Bounce
5. Storytelling

The following sections look at these in more detail and provide scaffolding suggestions you can use for Fatou at the end of each strategy.

PICTURE-WORD MATCH

Greet your students at the door and hand them an index card that either has a vocabulary word on the card (written in English as well as the languages of the students in your class), or a picture representation of a vocabulary word. Instruct students to move around the room and find the person who has the complementary card (the vocabulary word to match the picture or vice versa). When all the students find their match, have them discuss with each other why they feel they are a pair.

Figure 5.2 shows an example of a matched pair.

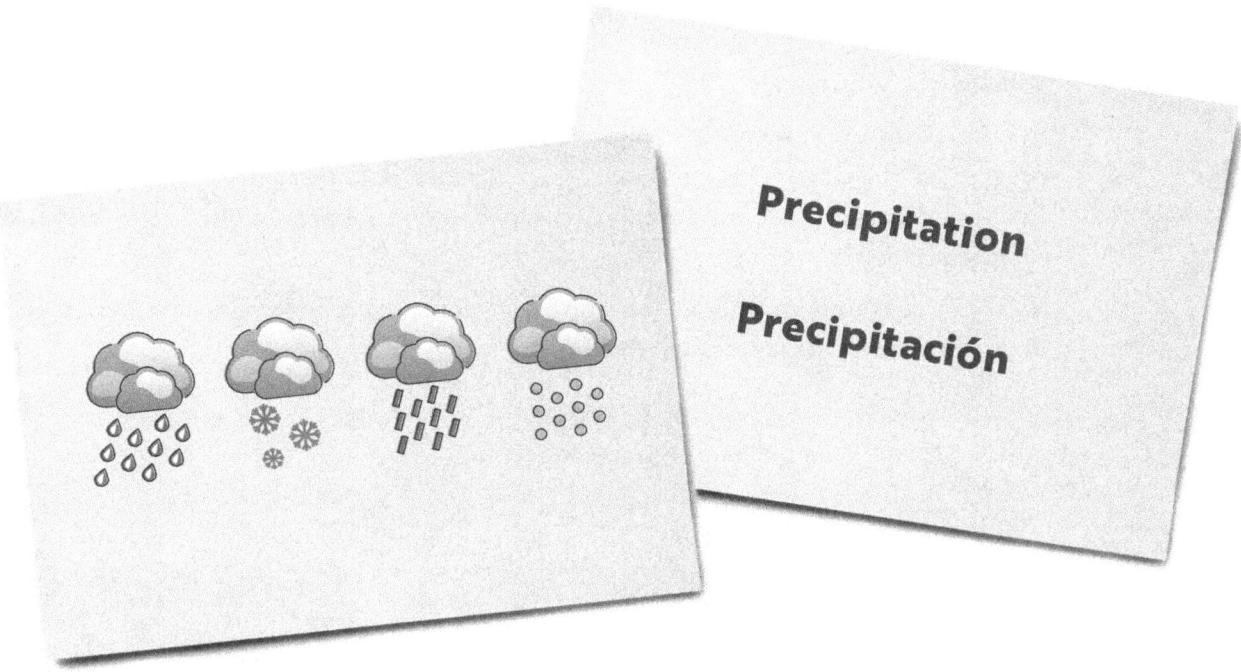

FIGURE 5.2: Sample picture-word match.

Here are two scaffolding ideas you can use with Fatou for this strategy.

1. You may want to make sure that Fatou, whose language proficiency is between levels 1 and 2, receives the picture card. The student who has the written word card has a higher level of language proficiency or is a native speaker. If possible, make sure that Fatou is paired with a person who speaks the same language.

2. Provide Fatou with sentence frames like the following.

    * My card is a picture of . . .
    * Your card has the word _____, which matches my picture
    * Our pairs match because . . .

While all of this is occurring, walk around the room listening to the students converse and interact with each other while assessing their knowledge. This will help you know which students, including Fatou, will need more scaffolds and assistance as you teach the lesson. Once the students have decided on their matches and discussed them with each other, you can project the correct answers on the board and review the vocabulary words that the students will be learning. This is *not* the time to teach vocabulary. You are presenting the vocabulary words to ascertain their preexisting degree of knowledge. Remember that the brain wants to make sense of the information before the facts. Allowing students the opportunity to make sense of the word and picture matches first provides them with the chance to connect the word to their cognitive schema. *Cognitive schema* is a packet of information in the brain that categorizes objects and concepts into groups. Having schemata in our minds make it easier to identify new objects and try to define them based on our existing knowledge of similar objects and concepts (Cornell, 2024).

THE SLOW REVEAL

This warm-up activity uses a picture that is slowly revealed to the students in separate segments. There are three parts or photos, with each stage revealing more information. At each step, students list or state things they see and make predictions about what the full image will be.

Figure 5.3 shows an example using an image of the water cycle, which is revealed in three stages, with each stage revealing more of the water cycle process.

Picture 1

What do you notice?
What do you see?

Picture 2

What more do you see now?
What do you think this is a picture of?

Picture 3

What is this? Have you seen this before? What questions do you have?

FIGURE 5.3: Sample slow reveal activity.

Students list the things they see and make predictions about what it is. Let students use the language they are most comfortable with—this activity's purpose is to determine their background knowledge in the topic. List students' ideas on the board and use prompts or clarifying questions. By having students list these ideas, you can use a translation app to translate what they have written. On the final image reveal, ask the students questions, such as "Who has seen this before?," "What does it represent?," and so on, drawing attention to the arrows and their directions or whatever background information you believe students need to know.

Allow students to explain what they know about the picture or vocabulary words in small groups then with the whole group, keeping in mind the characteristics of what each student can do at each level of language proficiency. Then, connect the picture to the unit and explain what they will be learning about. Finally, ask a perplexing question that introduces the STEM challenge. In this case the perplexing question would be, "Does it rain the same amount in every city?"

Here are two scaffolding ideas you can use with this strategy.

1. Show Fatou and the class a time-lapse video of the water cycle or videos of water evaporating. Ask students what they notice about the water during each phase. To connect this to Sudan, Fatou's home country, you may want to show the rainy season in Juba or the Congo—the rainiest area—and then its dry season, which is from December to February, to show how the water dried up, and connect it to the water cycle.

2. Ask the student if it rains a lot where they are from. This question is also relevant to multilingual learners born in the United States. Once again you can have pictures of Sudan during its rainy season for Fatou to share with the class.

FLEXIBLE VOCABULARY JOURNAL

A flexible vocabulary journal is not a journal where students write definitions in a notebook. It is a journal with three distinct columns (left, middle, and right) that allow students to indicate words they know, words they've heard of, and words they don't know. Students' goal is to move the vocabulary words to the left column (words they know with certainty) by the end of the unit. This word movement makes the journal flexible. Use the following steps to complete the activity.

1. Provide students with the list of vocabulary words, including pictorial representation, that you'll be teaching during the lesson.

    a. *evaporation*

    b. *condensation*

    c. *precipitation*

    d. *states of matter*

5. Provide each student with the flexible vocabulary chart as seen in figure 5.4.

6. Instruct students to chorally recite the vocabulary words.

7. Cue students to discuss the science words in pairs and complete individual charts by placing the words in the appropriate column based on their knowledge. Alternatively, have students sort the words into two or more separate categories and create a title for each category.

8. As you teach the lesson, ask students to self-evaluate their knowledge of the word or words before, during, and after instruction. Ask students if they have any vocabulary words that can be moved left to a new column.

9. The goal is for all words to be in the first column by the end of the lesson.

Goal: Move words left ←

Words you know with certainty	Words you've heard of, but are not sure of the meaning	Words you do not know or have not seen

FIGURE 5.4: Flexible vocabulary journal.

Visit **go.SolutionTree.com/EL** *for a free reproducible version of this figure.*

As students are working through the lesson, keep an eye on their charts to ensure words are moving to the left. If that is not occurring for any student, you will need to intervene to help the student learn the new words. This is another time where Two Pairs in a Quad can be beneficial.

Here are two scaffolding ideas for Fatou you can use with this strategy.

1. Allow the other students in the quad to assist Fatou with understanding the word. All students benefit from learning critical-thinking skills and learning strategies that are used naturally by the highest-performing multilingual learners (Tharp, 2000; Zohar & Dori, 2003).

2. Use the Frayer Model, a four-quadrant graphic organizer that helps students understand vocabulary or concepts by defining the term, listing its characteristics, and providing examples and nonexamples. The model promotes deeper comprehension by encouraging critical analysis and differentiation of related concepts or similar framework to help students understand the word. You can use categories, such as illustrations, symbols, synonyms and antonyms, foreign language translations, model sentences, and nonexamples, to assist Fatou in completing the Frayer Model and help her understand the vocabulary word.

POSE, PAUSE, POUNCE, AND BOUNCE

This strategy is more about assessing and building background knowledge than teaching vocabulary. In the water cycle example, initiate the lesson conversation using the term *rain* as the avenue to engage the students, since most, if not all, kids have experienced rain. Begin the dialogue with one student and bounce to another by taking the statement of the first student and selecting another student to elaborate on it or add to what the previous student said. This type of questioning is called Pose, Pause, Pounce, and Bounce.

Let's see how this might play out in the following interaction between a teacher and a few students, including Fatou. In this example, the teacher uses a thumbs-up or thumbs-down strategy to get a quick view of student responses. The teacher asks one question at a time, includes visuals, enunciates, and provides enough wait time for the multilingual learners in the class to process the question before asking for their reaction.

> *Teacher:* So, have all of us seen or felt rain? What about downpours? [Uses visuals for Fatou.]
>
> *Rafael:* My family lived in the tropics. We would get a lot of rain. Sometimes, there would be so much rain, it would look like rivers running down the streets.
>
> *Fatou:* [Uses a thumbs-up to represent her answer.]
>
> *Teacher:* Has anyone else experienced this type of heavy rainfall? [Bounces this question to specific students and listens to their responses.] Fatou and Bremen?
>
> *Bremen:* Yes. I have also seen this, especially during a hurricane. There is a lot of rain, and it is very scary.
>
> *Fatou:* Yes. We have time with much rain.
>
> *Teacher:* But after the downpour, does the rain remain on the streets and roads forever? Yes or no?
>
> *Lindsey:* No. The roads dry up.
>
> *Fatou:* No.
>
> *Teacher:* [Bounces to another student with a follow-up question.] Pieter, do you notice this as well? Tell me what you notice about the weather when the roads dry up. Is it sunny? Is it cloudy? [Simplifies the question for Fatou.] Fatou, when the rain goes away, what does the sky look like? [Shows road with rain, then no rain, and each time points to the road and the sky.]
>
> *Fatou:* [Points to the wet road picture.] The rain picture has a cloudy sky. [Points to the dry road picture.] And the dry road has a sunny sky.
>
> *Pieter:* I notice that when the sun comes up, the roads dry up faster than when there is no sun.
>
> *Teacher:* [Restates what each student said before continuing the lesson.] Well, in science, the rain drying up is called evaporation. [Shows picture.] The water from the rain does not just simply disappear. It changes form due to the heat from the sun. [Introduces states of matter.] It causes the water that was on the road and grass

to evaporate. It also causes water from the oceans, lakes, and streams to evaporate as well. Evaporation occurs when liquid turns to gas.

You can also use this strategy when introducing the other phases of the water cycle. The key is to use something all students can relate to or have experienced in order to build background knowledge. The students will be more engaged if they can relate, and even students at lower levels of language proficiency can connect to the content.

Here are two scaffolding ideas you can use with this strategy to assist Fatou in understanding more challenging vocabulary—in this case, *evaporation*.

1. Use a hot plate and place a small amount of water in a beaker on it. Turn the heat on and have students watch as the water gets less and less and water droplets form on the inside of the beaker until the water is gone. Have the students draw a picture of what they observed from the beginning, middle, and end. You want the students to show that the water level decreased. Then ask, waiting for responses from the native English-speaking students, as well as Fatou, "Where did the water go?" [*Shrug your shoulders in bewilderment.*] "Did I pour the water out of the beaker?" [*Shake your head in the motion for a yes or a no.*] "You have just witnessed *evaporation*!" [*Write that word on the board and repeat the word multiple times.*]

2. Show a time-lapse video of evaporation occurring. Or more dramatically, how a body of water dries up. Think of the dry season in Africa, when the water dries up and all that is left is cracked, dried dirt.

STORYTELLING

Students who are learning English have their own unique sets of experiences and backgrounds that may be very different from their native-speaking peers. Bridging the connection from what they already know to what we want them to know will help them master content more easily (Continental Press, 2022). Storytelling can be that bridge. It's an effective strategy because it enhances listening skills, increases focus and concentration, teaches vocabulary, and promotes empathy. Digital storytelling has been found to have an effect size of 0.714 in elementary, 1.092 in middle school, and 0.859 in high school (Akgün & Akgün, 2020).

When the story becomes a personal narrative, it takes on a life of its own as an effective tool that incorporates both literacy and cultural knowledge (Hill, 2023). Abeer Shinnawi (2021), program lead for Reimagining Migration, writes that storytelling is a powerful tool in a safe learning environment with classroom norms that reflect respect, honor, and empathy. When students can tell their own stories, it helps reverse a deficit mindset to create an asset mindset focused on what immigrant students can offer in each classroom (Shinnawi, 2021). But reading aloud also has clear cognitive benefits. According to education journalist Deborah Farmer Kris (2018), "brain scans show that hearing stories strengthens the part of the brain associated with visual imagery, story comprehension, and word meaning. One study found that kindergarten children who were read to at least three times a week had a 'significantly greater phonemic awareness than did children who were read to less often.'" So, storytelling, whether from a book or from a personal perspective, is an effective strategy to use with your multilingual learners.

Storytelling can take two basic formats: (1) using children's picture books or children's short-story books to connect the story to the content or (2) storytelling from the perspective of sharing personal experiences. Picture books can be particularly useful for helping multilingual learners understand abstract concepts, such as emotions or cultural values. They can also be useful in helping students understand conceptual ideas in other content areas, such as mathematics or social studies (Continental Press, 2022). Storytelling from the perspective of sharing personal experiences can be a powerful and authentic way to build a student's skills and to identify their funds of knowledge on a specific topic (Hill, 2023). Scan the QR code to view a short video on the power of using storytelling in the classroom and its connection to the brain.

I often used picture books when I taught at the elementary, middle school, and high school levels. Many may think that high school students would not be interested in having their teacher read them a children's picture book, but they were, and are, interested. It all depends on the delivery. I always start by asking, "Do you think this book can teach you, a high school student, about a high school science topic?" For elementary-aged students, I ask, "What do you think this book is about?" I show the cover of the book, and follow up by asking, "How do you think it can be used for science?"

Storytelling and the Brain
https://scanqr.to/08a51f61

Beginning a lesson with a picture book can help students conceptualize the content, make sense of it prior to the teacher providing scientific facts and vocabulary, and provide the teacher with the degree of students' funds of knowledge on this principle.

Before you read the book, task your students with the following.

- Have the students write down what they notice and wonder as they listen to the story. For younger students who may not be able to write well yet or for the multilingual learner at a proficiency level of 1 or 2, you can stop along the way and ask them what they notice and what they wonder. Remember that level 1 will either provide a one- to two-word answer or not participate, so think of yes-or-no questions to ask.

After you finish reading the book, task the students with the following.

- With their quad, select three things they notice and three things they wonder that they want to present to the class.
- Have one person from each quad write their three notices and wonders on their table's whiteboard.

Discuss each one as a class and eliminate redundant responses from other quads. As you read the book, check to see if any students are jotting things down. If students seem reluctant, tell them what you notice and what you wonder, and write them on the classroom whiteboard. Then a few minutes later, ask them what they notice and wonder and instruct them to write it on their paper (multilingual students can have their quad assist them in completing the exercise). Do this throughout the book so they understand the process and expectation. Then, ask questions to get an understanding of what they know (their funds of knowledge). This will help you decide the depth and breadth that you should teach for the lesson and give ideas on how to connect the concept to student responses.

Let's look at how a teacher might use the book *Archimedes and the Golden Crown* (Chowdhury, n.d.) when teaching the concept of Archimedes' principle, and the concept of using water to determine the volume of an irregularly shaped object, to middle and high school students. For Fatou, the teacher provides a sheet of paper with sentence starters. For example, I notice _____; I wonder why _____. The teacher also provides Fatou with visuals for words that she may have difficulty understanding, and polysemous words, such as volume for sound, and volume for the amount of space an object occupies.

> ***Teacher:*** *Hey, everyone, I want to start the lesson differently today. I am going to read to you a short book called, Archimedes and the Golden Crown. [Shows the book to the class while pointing to each word in the title to assist Fatou in understanding.] Has anyone ever read or heard of this book? [Teacher nods head in a yes motion and says "yes" and shakes head in a no motion and says, "no."]*
>
> ***Class:*** *[A few students shake their heads to indicate they haven't heard of the book.]*
>
> ***Teacher:*** *OK. I am going to read the book and I want you to jot down what you notice and what you wonder, so be sure to have a pencil or pen and the worksheet in front of you. [The teacher does not call Fatou out in front of the class. The teacher created a worksheet for the class with an image of an eye for what students notice and an image of a question mark for what they wonder. The teacher models the action for the class using the worksheet by projecting it on the board and modeling how to complete it.]*
>
> ***Student:*** *Seriously? You are going to read us a kid's book?*
>
> ***Teacher:*** *Yes! And it connects to our lesson. Are you all willing to give it a try?*
>
> ***Class*** **[with some grunts]*:*** *OK. [Fatou gives a thumbs-up.]*
>
> *[The teacher initiates class discussion after reading the book to ensure that students understand the point of the story and the connection to Archimedes' principle.]*
>
> ***Teacher:*** *What was Archimedes tasked to do? [Shows the page where the king asks Archimedes to complete the task for him.]*
>
> ***Student:*** *Create a new crown for the king made of gold.*
>
> ***Teacher:*** *Discuss the story with your quad. Think about the following question in your discussion: What happened when Archimedes sat in the tub? [The teacher leads the students to stating that the water overflowed and walks around, listening to the conversations and asking additional questions. The teacher stops to listen to Fatou's quad, comprising Ella (H), Max (MH), Sam (ML), Fatou (L).]*
>
> ***Ella:*** *OK, so we need to talk about the book Archimedes and the Golden Crown. Did everyone understand the story about how he figured out the crown wasn't pure gold?*
>
> ***Max:*** *Yeah, it was pretty cool! He got into the bath and noticed the water rise, and that's how he figured out how to measure the crown, right?*

Sam: Exactly! He used the water to measure how much the crown displaced the water, which told him about its density. That way, he could tell if it was real gold or mixed with something else. Fatou did you understand the story? It's like when you drop something in a bucket full of water, the water goes up. So, Archimedes used that idea to catch the king's crown maker cheating.

Fatou: [Nods her head to state that she understands.]

Ella: Right, Sam! He was so excited when he figured it out that he ran through the streets yelling "Eureka!" which means "I have found it!" in Greek.

Max: That's so funny! Imagine running outside and yelling because you're so happy about your homework.

Sam: And it's not just about the crown. This story shows how smart Archimedes was. He invented a lot of other things too.

Fatou: This story see him as smart. He helped the king.

Teacher: [Calls the class back together.] How many of you take showers or baths?

[The teacher asks Fatou if, in her country, they use a bathtub or a shower to bathe and presents pictures of each. The teacher states that in many countries, there are only showers. Following, the teacher asks the class if this was the case, what would they have done? The teacher then states that at the base of the shower there may be something to keep the water from flowing all over the floor.]

Teacher: How would this have impacted the story?

For a visual, place a cup filled to the brim with water on each quad table along with a small plastic toy or wooden block. Ask students to drop the toy or block into the cup and discuss with their table what happened. This visual mimics Archimedes sitting in the tub and water overflowing to the floor. Have the students write what they notice as they drop the object into the cup filled with water. They should do this with different-sized objects each time with a cup filled to the brim. The objective is to see if students can make any connections between the amount of water (volume) that overflows and the size of the object. Students should notice that the larger the object, the more water that overflows. This connects to the volume of the object.

Another option is to have the students drop rectangular or square objects into the water. Use rectangular or square objects because it is simple to calculate their volume using the equation length × width × height. Have the students compare their calculated answer to the volume of water that overflowed. The students should notice that the object's calculated volume and the volume of the water overflow are the same or very close. Then, have students discuss how they could manually calculate the volume of an irregularly shaped object, like a crown (this links the activity back to the book). After their discussion, ask if there is an easier way to determine the volume using water. The point is for the students to draw the conclusion that water overflow, like the overflow from Archimedes' tub, can be used to easily determine the volume of an irregularly shaped object.

This approach allows students to make sense of the situation before giving details. This is a brain science fact! Remember, the brain wants to make sense of a situation before it wants details. To be exact, if you want students to pay attention, don't start with details; start with the key ideas and, in a hierarchical fashion, form the details around these larger notions (Medina, 2014). In other words, allow students to discover and make sense *their way* first before you define it or give a lot of scientific facts and details. By using hands-on, concrete objects, you have removed the barrier of language to demonstrate the concept. Now, the vocabulary can be defined by the association to the activity.

The STEM Challenge—Introduction

Now that you know how to connect students' assets to instruction, activate their background knowledge, and utilize scaffolded strategies to accelerate their learning, it's time to put everything together as you plan your STEM challenge. The Global Precipitation Measurement Mission (GPMM) challenge: Precipitation Towers (GPMM, n.d.b) has two parts and fulfills the following basic suggestions.

- Provides students voice and choice (figure 5.1, page 122) by allowing the following.
    - For grades K–5, the students choose from preselected cities with the precipitation data in whole numbers and decimals.
    - For grades 6–12, the students choose cities or countries from around the world; they must research their city or country and obtain the precipitation data themselves.
- All students have had some interaction with weather and various types of landforms, so the topic will not be foreign or hard to grasp (compared to a topic like developing a better football helmet that will lessen the incidence rate of concussions, since not all students have experience with American football).
- The challenge is very hands-on and provides many visuals to aid students in making sense of what is happening without having to rely entirely on academic vocabulary to formulate a basal level of understanding.
- The challenge uses scaffolding that allows students to systemically build their knowledge base and gain confidence when performing tasks independently.

The parts of this STEM challenge are as follows.

1. **Precipitation Towers:** This section uses stacking cubes to graph precipitation data and compare the precipitation averages and seasonal patterns for several locations. This activity has variations that can accommodate various ages and ability levels. For grades K–5, the activity comes with three levels to explore weather patterns and precipitation averages: (1) beginner (whole numbers for precipitation amounts), (2) intermediate (decimals for precipitation amounts), and (3) advanced (the students must research the location and complete the table on their own).

2. **Global Precipitation Measurement Mission:** This section teaches students about how geographical features impact weather and climate, how to graph data, how use that data to create a climatograph, and how to draw conclusions based on the data. For grades 4–12, you will see how scaffolding is integrated into the lesson, allowing you to seamlessly assess student knowledge.

Figure 5.5 outlines the Next Generation Science Standards (NGSS) and the Common Core Mathematics Standards for the STEM challenge.

STEM challenge overview:	Students will research precipitation data for a city of their choosing; graph data using stacking cubes (visual) or by manually creating a graph; compare the precipitation averages and seasonal patterns; and apply their knowledge of the water cycle to help define keywords, such as precipitation, geography, mountains, rain shadow, climatogram, elevation, water cycle, evaporation, condensation, and weather.
NGSS Evidence Statement	Develop a model to describe the cycling of water through Earth's systems driven by energy from the sun and the force of gravity.
Science and Engineering Practices	Developing and Using Models • Use observations (firsthand or from media) to describe patterns in the natural world to answer scientific questions. (K-ESS2-1)* • Represent data in tables and various graphical displays (bar graphs and pictographs) to reveal patterns that indicate relationships. (3-ESS2-1)* • Analyzing and Interpreting Data • Develop and use a model to describe phenomena. (MS-ESS2-6)
Crosscutting Concepts	Patterns, Systems, and System Models Patterns • Patterns in the natural world can be observed, used to describe phenomena, and used as evidence. (K-ESS2-1)* • Patterns of change can be used to make predictions. (3-ESS2-1), (3-ESS2-2)* Systems and System Models • A system can be described in terms of its components and their interactions. (5-ESS2-1)* • Models can be used to represent systems and their interactions—such as inputs, processes, and outputs—and energy, matter, and information flows within systems. (MS-ESS2-6)
Disciplinary Core Idea (DCI)	MS-ESS2-D: Weather and Climate Weather is the combination of sunlight, wind, snow or rain, and temperature in a particular region at a particular time. People measure these conditions to describe and record the weather and to notice patterns over time. (K-ESS2-1)* • Scientists record weather patterns across different times and areas so they can make predictions about what kind of weather might happen next. (3-ESS2-1)* • Climate describes a range of an area's typical weather conditions and the extent to which those conditions vary over the years. (3-ESS2-2)* ESS2.A: Earth Materials and Systems • Winds and clouds in the atmosphere interact with the landforms to determine patterns of weather. (5-ESS2-1)*

*The lower grade levels are provided if you want to know where those skills are first introduced to assist your multilingual learners in understanding the concepts.

Source: NASA Jet Propulsion Laboratory (n.d.) and PBS LearningMedia (n.d.).

FIGURE 5.5: STEM challenge overview.

The NGSS evidence statement provides a deeply detailed outline of what students should be able to do at the end of learning this science standard. These standards are developed through productive application, which are observable by educators. The NGSS framework can support language development lessons because it encourages students to discuss their experiences, observations, and conclusions as they work through the scientific process. The NGSS also establishes language and science domain skills simultaneously that support success in the language classroom and beyond (Davila, 2023).

Figure 5.6 shows the mathematical standards.

Common Core Mathematics Standards (CCSS)	
Standards for mathematical practices (MP) For this STEM Challenge: MP2: Reason abstractly and quantitatively MP4: Model with Mathematics	Have students calculate the range of their data by subtracting the lowest amount of precipitation from the highest. If appropriate, have them calculate the annual precipitation averages for their location. The eight standards for mathematical practices describe varieties of expertise that mathematics educators at all levels should seek to develop in their students. These practices rest on important processes and proficiencies with long-standing importance in mathematics education.
K.MD.A.2	Directly compare two objects with a measurable attribute in common, to see which object has "more of" or "less of" the attribute and describe the difference.
1.MD.A.2	Express the length of an object as a whole number of length units, by laying multiple copies of a shorter object (the length unit) end to end; understand that the length measurement of an object is the number of same-size length units that span it with no gaps or overlaps. Limit to contexts where the object being measured is spanned by a whole number of length units with no gaps or overlaps.
1.MD.C	Organize, represent, and interpret data with up to three categories; ask and answer questions about the total number of data points, how many in each category, and how many more or less are in one category than in another.
2.MD.D.10	Draw a picture graph and a bar graph (with single-unit scale) to represent a data set with up to four categories. Solve simple put-together, take-apart, and compare problems using information presented in a bar graph.
3.MD.A	Solve problems involving measurement and estimation.
3.MD.B.3	Draw a scaled picture graph and a scaled bar graph to represent a data set with several categories. Solve one- and two-step "how many more" and "how many less" problems using information presented in scaled bar graphs.
4.MD.B.4	Make a line plot to display a data set of measurements in fractions of a unit ($\frac{1}{2}, \frac{1}{4}, \frac{1}{8}$). Solve problems involving addition and subtraction of fractions by using information presented in line plots.

5.MD.B.2	Make a line plot to display a data set of measurements in fractions of a unit ($\frac{1}{2}, \frac{1}{4}, \frac{1}{8}$). Use operations on fractions for this grade to solve problems involving information presented in line plots.
5.NBT.A.4	Use place value understanding to round decimals to any place.
6.SP.B.5	Summarize numerical data sets in relation to their context.
7.SP.A.1	Understand that statistics can be used to gain information about a population by examining a sample of the population; generalizations about a population from a sample are valid only if the sample is representative of that population. Understand that random sampling tends to produce representative samples and support valid inferences.
8.SP.A.1	Construct and interpret scatter plots for bivariate measurement data to investigate patterns of association between two quantities. Describe patterns such as clustering, outliers, positive or negative association, linear association, and nonlinear association.
HSS.ID.A.1	Interpreting categorical and quantitative data

Source for standard: National Governors Association Center for Best Practices & Council of Chief State School Officers, 2010.

FIGURE 5.6: Mathematical standards.

The Common Core mathematics standards concentrate on a clear set of mathematics skills and concepts. Students will learn concepts in a more organized way both during the school year and across grades. The standards encourage students to solve real-world problems. But that is not always enough to ensure multilingual learners' academic success. Their level of academic language proficiency makes acquiring and assessing content-area knowledge challenging (Francis, 2006). Therefore, scaffolds must still be embedded into instruction so that multilingual learners have the most opportunity for success. But these learners, like any other population of learners with academic difficulties, require effective instructional approaches and interventions to prevent further difficulties and to augment and support their academic development (Francis, 2006).

I have found that preparing what I refer to as a STEM minilesson (think chunking) to be effective when working with multilingual learners. These minilessons contain much of the knowledge and skills that are required for the STEM challenge. A STEM minilesson provides the following benefits: (1) offers the teacher insight into the student's background knowledge, and (2) gives the student background knowledge in the event they don't have experience with the content.

Figure 5.7 (page 138) compares the objectives of the STEM minilesson and the STEM challenge so you can see how they complement each other and build and develop knowledge and skills that multilingual learners need for the STEM challenge (GPMM, n.d.a).

STEM Minilesson Objective	STEM Challenge Objective
•Students brainstorm about geographic features, consider how they might affect temperature and precipitation, and discuss the difference between weather and climate. •Students develop key vocabulary related to weather and precipitation.	•Students examine data about a location and calculate averages and compare with other locations to determine the effect of geographic features on temperature and precipitation. •Students use stacking cubes or create bar graphs to represent precipitation data. •Students develop key vocabulary related to weather and precipitation and understand how the sun and energy impact weather. •Students compare temperature and precipitation graphs for various locations to look for patterns in geographical influence on climate, then collect data for a location of their choice and create their own climatogram.

FIGURE 5.7: Comparing objectives of the minilesson and the STEM challenge.

The lesson plan template for the STEM minilesson and the STEM challenge will use the 5E inquiry-based instructional model, which we'll call the 5Es for short. Educator and author Sam Northern (2019) explains that the 5Es are based on cognitive psychology and constructivism, which postulates that learners build or construct new ideas on top of their old ideas and best practices in STEM instruction. The 5E learning cycle leads students through five phases: (1) engage, (2) explore, (3) explain, (4) elaborate, and (5) evaluate. The model brings coherence to different teaching strategies, provides connections among educational activities, and helps science teachers make decisions about interactions with students. Compared to traditional teaching models, the 5E learning cycle results in greater benefits concerning students' ability for scientific inquiry. In other words, the 5Es are the guide to engaging students in scientific inquiry. Teachers aiming to engage students, motivate them to learn, and guide them toward skill development will use the 5Es for this purpose (Duran & Duran, 2004; Lesley University, n.d.).

Another effective benefit of using the 5Es is that they mirror the way scientists really approach their work. They start with a question, investigate, form an explanation, test that explanation in new contexts, and then evaluate the results. The 5E model also prioritizes student engagement and ownership of learning, which can be more effective than traditional, lecture-based teaching methods. Rodger W. Bybee, co-creator of the 5E model, explains how to make the most of the model (Lesley University, n.d.):

> The 5E Model is best used in a unit of two to three weeks in which each phase is the basis for one or more distinct lessons. Using the 5Es model as the basis for a single lesson decreases the effectiveness of the individual phases due to shortening the time and opportunities for challenging and restructuring of concepts and abilities—for learning. And if too much time is spent on each phase, the structure isn't as effective and students may forget what they've learned.

Table 5.1 describes the 5E phases and how to implement them in your instruction.

TABLE 5.1: 5E Phases

The 5E Phases	Description	Implementation
Engagement	This first step of the 5Es is mainly to stimulate curiosity, activate prior knowledge, and foster an interest in upcoming concepts so students will be ready to learn (Lesley University, n.d.). The teacher works to gain an understanding of the students' prior knowledge, identify any knowledge gaps, and identify possible misconceptions (Northern, 2019). The teacher presents a problem, asks a question, or shares an interesting phenomenon to captivate students' attention. This is not the time for the teacher to lecture, define terms, or provide explanations.	This phase should use picture observations, video clips, demonstrations, kinesthetic activities, a discussion of students' prior experiences, a provocative question, or a free write (students write continuously without stopping; teachers usually use a timer to provide limits).
Explore	This second step of the 5Es is when the teacher provides students with hands-on activities that will help them use prior knowledge to inquire, generate new ideas, and conduct a preliminary investigation. Students actively explore their environment or manipulate materials. It's essential that students have time to think, plan, investigate, and organize collected information. The goal is to get students directly involved and interested in the topic, enabling them to ask questions, hypothesize, and make predictions.	This phase is where students will ask a testable question, conduct research, and form a hypothesis. Then, they will test the hypothesis and gather data, analyze and interpret the data, draw a conclusion, and communicate results.
Explain	This is a teacher-led phase that helps students synthesize new knowledge and ask questions if they need further clarification (Lesley University, n.d.). This phase gives students an opportunity to communicate what they have learned and form their understanding into a coherent explanation. Teachers ensure that students' understanding aligns with the scientific explanation by asking students to share what they learned during the explore phase before introducing technical information in a more direct manner.	This phase should involve activities like discussions, presentations, or the introduction of formal terms and definitions.
Elaborate	In this phase, students are encouraged to apply what they have learned to new situations, broadening, and deepening their understanding. The main objective is to reinforce the main concept, enhance comprehension, and allow students to make connections to other related concepts.	This phase might involve new problems to solve, projects to tackle, or deeper investigations related to the initial topic. In this phase, students cement their knowledge before being evaluated.
Evaluate	The final phase focuses on both formal and informal assessments. Teachers can observe their students and determine, by listening to conversations or reviewing assessments, if students have completely grasped the core concepts.	This phase can be completed through tests, quizzes, self-assessments, peer assessments, writing assignments, exams, oral discussions, or presentations. It's important to note that while this phase is listed last, evaluation should be ongoing throughout the entire 5E cycle.

The 5Es instructional model is a teaching framework that helps guide students through the learning process in a structured way. It benefits multilingual learners by providing a clear and structured learning process that supports language development and comprehension. This model's emphasis on interaction, application, and assessment helps multilingual learners develop both their academic and language abilities effectively and helps students build a strong foundation of knowledge and skills.

Two tools in this lesson that facilitate learning for multilingual learners are chalk talk and inquiry charts. *Chalk talk* is a visual method of teaching where the teacher uses a chalkboard, whiteboard, or posterboard to visually present ideas, concepts, and information to students either among student groups or as a whole class. Consider the following benefits multilingual learners receive from chalk talk. Each item includes instructions for teachers.

- Vocabulary development
  - Use the chalkboard, whiteboard, or posterboard to write down new vocabulary words alongside images or drawings that represent their meaning. This can help multilingual learners make connections between the new words and their meanings more easily.
  - Write synonyms, antonyms, or sentences using the new vocabulary to provide context.
- Visual storytelling
  - Illustrate simple stories or scenarios on the board while narrating them. This can help multilingual learners understand the narrative while associating new vocabulary words and expressions with the visual context.
- Concept mapping
  - Create visual organizers or concept maps on the board to show relationships between ideas, or to organize thoughts for writing assignments.
- Error correction
  - Use the board to correct common mistakes made by students in a visual way, which can help to reinforce the correct usage or structure.
- Discussion and feedback
  - Write questions or topics on the board to prompt discussion.
  - Use the board to jot down points made during discussions or to provide feedback on student responses.

Inquiry charts are graphic organizers that are similar to K-W-L charts (know, want to know, learned) that allow students to gather information about a topic from several resources. Because both types of charts access background knowledge, this strategy helps students integrate ideas from several sources (Wisconsin Department of Public Instruction, n.d.). As one begins using inquiry charts, the teacher can select the topic and provide inquiry questions and sources. More advanced students may select their own topics, prepare their own inquiry questions, and find their own sources (Ullman-Shade, 2015). The inquiry chart is a GLAD strategy. To learn more about GLAD strategies, visit https://ntcprojectglad.com.

The visual aspect of inquiry charts and chalk talk can be very beneficial for multilingual learners as it provides a visual aid to understanding the material, which can be particularly helpful when learning a new language. These tools can help students identify inaccurate information as well as connections and similarities between their knowledge. They are also an avenue for discussion among students and the whole class.

Consider the following benefits multilingual learners receive from inquiry charts.

- Teaches students to predict based on their personal and background knowledge.
- Creates opportunities for academic discourse and comprehensible output.
- Sets purpose for learning and creates opportunities for academic discourse.
- Preassesses concepts and oral language.
- Assesses vocabulary, grammar, English language development level, and language production.
- Creates opportunity for modeled writing.
- Introduces hypothesis model.
- Lowers affective filter, all contributions charted.

To prepare to use chalk talk and inquiry charts, decide what topics and inquiry questions will be explored, decide what sources will be used, and provide students with the graphic organizer.

To help multilingual learners participate in chalk talks and inquiry charts, use the following sentence frames for language acquisition:

- What I know about _____ is _____.
- Based on _____ I know about _____.
- I have a question about _____.
- I am beginning to learn _____.
- I learned _____.

Figure 5.8 (page 142) contains the lesson plan for the minilesson. Remember that this minilesson provides an opportunity to assess and build on Fatou's background knowledge of weather. It gives her a chance to make sense of vocabulary words involved in the STEM challenge and work with peers to create a climatogram.

STEM Minilesson: Preparation for Global Precipitation Measurement Mission
Objective: Determine students' funds of knowledge regarding the effect of geographic features on weather and climate. Introduce students to inquiry charts, chalk talk, and vocabulary.
Grade Level(s): 4–8; middle school*

5E Lesson Plan	Activity (What Teacher Must Do)
Engage Time required: One class period You will need: Video or pictures of various weather conditions based on geographic features; NASA's Precipitation Education page has a plethora of videos, images, and articles to choose from (NASA, n.d.a)	Teacher will: Show students videos of various weather conditions based on geographic features. Have the students turn and talk about what they notice and wonder. Be sure to walk around and listen to the conversations. Listen for key precipitation words like rain, sleet, snow, or hail that are connected to the lesson or for common misconceptions. Have students share if they have experienced various weather conditions. For a cultural exchange, ask Fatou and other students who may be from other states or countries, or have family they have visited, to explain geographical features of where they lived and the weather there. Ask if they were near mountains, the ocean, a desert, and so on. Have students discuss what they noticed about the geographical features and the weather in the video or pictures you provided. Do they think that there is a connection?
Scaffolds for Fatou: If the video has narrations, see if you can provide Fatou with the written translation in advance in Arabic so she can read it before the STEM challenge begins. If not narrated, stop the video at various points and use simple sentences to point out features that connect to the vocabulary so she knows what she needs to focus on. Being able to make connections to and show pictures of Juba, where Fatou was born and lived, would be a great addition to the lesson and can allow Fatou to contribute to the lesson.	
Explore Time required: Half a class period to a full class period You will need: Chart paper, markers, visuals in the form of videos or pictures of various types of precipitation (rain, sleet, snow, or hail) and geographic features	Teacher will: Introduce and model how to complete the two strategies to be used in the STEM challenge (inquiry chart and chalk talk) so that students understand the expectation. The chalk talk and inquiry chart will have students demonstrate their understanding of geographical influences, focusing on key vocabulary words, such as precipitation, geographic features, measurement, data, climate, weather, climatogram, mountains, elevation, temperature, and graphing. You will use visuals, such as pictures, videos, and so on, to lay the foundation. Websites resources: https://oeta.pbslearningmedia.org/resource/buac17-68-sci-ess-global precipitation/global-precipitation https://gpm.nasa.gov/education/lesson-plans/geographical-influences
Scaffolds for Fatou: The teacher will use simple sentences and provide guided steps either by modeling or picture instructions to aid Fatou in completing the inquiry chart and chalk talk. Students will work in their Two Pairs in a Quad group to complete an inquiry chart and chalk talk. The teacher will allow time for students to practice in their quads while the teacher provides feedback.	

5E Lesson Plan	Activity (What Teacher Must Do)
Explain Time required: One class period You will need: Two-column chart (see figure 5.10, page 149), and the Venn diagram (see figure 5.11, page 150), pencils, sentence frames, Frayer Model or the word squares template	Teacher will: Write the two guiding questions on the board and also on a paper for each quad. Teacher will start a discussion with students based on the guiding questions and listen for understanding and misconceptions. Note that there are four guiding questions for the STEM challenge, but only two guiding questions in the STEM minilesson. The teacher will provide the students with the charts from figures 5.10 (page 149) and 5.11 (page 150). The students will turn and talk with their pair and complete the charts together. Once completed, the pairs will share their work as a quad, discuss, and make any changes to the charts based on their discussions. The teacher will also provide students with the Frayer Model or the word squares templates to complete for the vocabulary words. By using two of the four STEM challenge guiding questions for the STEM minilesson, Fatou and the students have time to make sense of the questions, hear them multiple times, discuss with their group and teacher, and understand the expectation on how to answer them and participate in the discussion. Guiding Questions: 1. Geographic features—What are geographic features, and how do you think they affect temperature and precipitation? Provide students with a two-column chart to complete (see figure 5.10, page 149). 2. Geographic features' influences on climate—How do you think nearness to large bodies of water, elevation, and the location of mountains affect temperature and precipitation? Provide students with images and a sentence frame to explain their answer (see figure 5.13, page 152). Finally, the teacher will review the answers with explanations using videos or pictures to reference and support the correct answers. The teacher will allow students to state their rationale as to why they selected the answers they did. This opens the door for external conversations and clarifications.

Scaffolds for Fatou:

The teacher should use simple sentences, visual aids, sentence frames, and realia to explain the concept to Fatou. This could include providing images of the vocabulary words alongside the words on a handout that she can cut out and paste on the chalk talk poster and add to it as she sees fit to show her understanding of the word. Allow Fatou the option to write in Juba Arabic to explain her understanding of the vocabulary word. Then, use a translation app to translate what she wrote to English so that you understand. Another option is to provide a Frayer Model or the word squares template and sentence starters for Fatou to complete.

Having the students work in their Two Pairs in a Quad will provide Fatou with the opportunity to draw or write what she knows about the topic in a smaller setting. This allows Fatou to make connections between her group's drawings or short sentences. It will also bring to light any of Fatou's misconceptions. This is the where the power of Two Pairs in a Quad becomes so important; her face and shoulder partners can be of great assistance in helping her participate and understand.

5E Lesson Plan	Activity (What Teacher Must Do)
Elaborate Time required: Half a class period to a full class period You will need: Climatogram from GPMM-NASA (visit **go.SolutionTree.com/EL** for a link)	Teacher will: Provide the students with two climatograms with different geographical features. The teacher will explain the completed climatogram to students and have them discuss, as a quad, what they notice about the weather patterns and climate.
Scaffolds for Fatou:	
The teacher should simplify the climatogram's language or translate. Use a city climatogram that has similar geographical features as Juba so that she can see the weather and climate similarities. Ask Fatou yes-or-no questions or have her point to answer your questions about the graph to check for understanding. However, since Fatou's language proficiency is at the speech emergence level, you can also ask her questions at a higher level than yes or no or pointing.	
Evaluate Time required: Fifteen to twenty minutes You will need: Climatogram	Teacher will: Provide students with a third climatogram to see if they can, as a quad, explain why the weather is like it is for that city. Since this is a minilesson, this should not be graded. Use this process to determine whether the students are ready to move on to the STEM challenge or if you must provide more scaffolds or assistance.
Scaffolds for Fatou:	
The teacher should provide translation as needed, use sentence frames, and let Fatou respond orally.	
Closure Time required: Fifteen minutes You will need: N/A	Inquiry chart: Have a class discussion about the inquiry charts and whether any students want to answer any of the questions posed or add questions. Chalk talk: Have students revisit the chalk talk board to comment on statements or questions. This is great for Fatou because she can interact in confidence with the activity.
Scaffolds for Fatou:	
The teacher should allow Fatou to use visuals to interact. Allowing Fatou to write in Juba Arabic and use a translation app to translate will help her be more engaged. If she is reluctant to speak in front of the whole class, make the closure occur in quads so that she can participate.	

FIGURE 5.8: STEM minilesson plan.

Now that Fatou and the students have completed the minilesson and you have been able to assess and build on Fatou's background knowledge of how geographical features impact weather and climate, it's time for the STEM challenge.

The STEM Challenge

Figure 5.9 contains the 5E STEM lesson plan.

STEM Challenge Lesson Purpose:

1. Students will compare precipitation data for a city of their choosing and graph that data using stacking cubes (visual) or by manually creating a bar graph (part 1).
2. Students will describe geographical features near the cities they chose in step one and compare graphs to find patterns in the effect of mountains, oceans, elevation, latitude, and so on, on temperature and precipitation (part 1 and part 2).
3. Students will research temperature and precipitation patterns at various locations around the world and use that information to create a climatogram (part 2).
4. Students will apply their knowledge of the weather and climate to help define keywords, such as precipitation, evaporation, condensation, weather, and climate change.

5E Lesson Plan	Activity (What Teacher Must Do)
Engage Time required: One class period Watch a video clip of global precipitation or view a map that shows average annual precipitation, in millimeters and inches for the world. Video clip (visit **go.SolutionTree.com/EL** for a link) Annual precipitation map (visit **go.SolutionTree.com/EL** for a link) For younger students: Show videos or pictures of different types of weather conditions and show where they are on a globe or map. Making references to geographical features and what they are.	Complete an inquiry chart and use chalk talk Topic question: (1) What factors influence weather? For Fatou: (1) What things affect the weather? Topic question: (2) What effect does geography have on climate? For Fatou: (2) How does the location of a place affect its weather over a long time? For younger students have them discuss using think-pair-share.

Scaffolds for Fatou:

You may need to discuss what a geographical feature is and use visuals, such as clip art, videos, and so on.

Fatou should understand what to do with the inquiry chart and chalk talk since it was presented in the STEM minilesson. Be sure her quad is still assisting her in the process.

As discussed in the STEM minilesson, if the video has narrations, try to provide Fatou with the Juba Arabic written translation in advance so she can read it before the STEM challenge begins. If it's not narrated, stop the video at various points and use simple sentences to point out features so she knows what she should focus on.

Being able to make connections to and show pictures of Juba, where Fatou was born and lived, would be a great addition to the lesson; this can allow her to contribute to the lesson and make her own connections.

5E Lesson Plan	Activity (What Teacher Must Do)
Explore Time required: One to two class periods You will need: Visual aids (images of cities, video or images of geographical features, different types of weather, precipitation, bar graphs, and so on, for association to the vocabulary words), printed vocabulary words, realia, sample city climatograms, poster board, Frayer Model, or the word squares template Geographical influences on climate (visit **go.SolutionTree.com/EL** for a link) Student capture sheet (visit **go.SolutionTree.com/EL** for a link) For younger students: 1. Let them know what geographical features exist for their city. 2. Determine what parts of the climatogram to point out based on your students' grade level and ability.	• Show a climatogram. Point out things like the title, axes, axes units, key, bar graph, and line graph. A climatogram is also known as a climate graph. It is a graphical representation of both the average monthly temperature and the average monthly precipitation of a specific region or location. • Discuss geographical influences, focusing on the key vocabulary words (precipitation, geographic features, measurement, data, climate, weather, climatogram, mountains, elevation, and temperature) and graph using chalk talk and inquiry charts. • Provide students with guiding questions (there are two additional questions not used during the STEM minilesson). Guiding Questions: • Geographic features: What are geographic features and how do they affect temperature and precipitation? Provide students with a two-column chart to complete (see figure 5.10, page 149). • Weather versus climate: What are the similarities and differences between weather and climate? Provide a Venn diagram for students to complete (see figure 5.11, page 150). • Climatograms: How can you easily see a location's climate data? What analysis tools can you use to compare the climate data for different locations? (see figure 5.12, page 151) • Geographical influences on climate: How do nearness to large bodies of water, elevation, and the location of mountains affect temperature and precipitation? Provide students with images and a sentence frame to explain their answer (see figure 5.13, page 152). Source for questions: Global Precipitation Measurement Mission, n.d.a.

Scaffolds for Fatou:

Be sure to use simple sentences, visual aids, sentence frames, and realia to explain concepts to Fatou.

Remember: Provide images of the vocabulary words alongside the words on a handout that Fatou can cut out, paste on the chalk talk poster, and add to as she sees fit to show her understanding of the word. Perhaps she can write in Juba Arabic to explain her understanding of the vocabulary word. Another option is to use a Frayer Model or the word squares template and sentence starters for Fatou to complete.

Having students work in their Two Pairs in a Quad will provide Fatou the opportunity to draw or write what she knows about the topic in a smaller setting. This allows her to make connections between her group's drawings or short sentences. It will also bring to light any misconceptions or inaccuracies Fatou may possess.

5E Lesson Plan	Activity (What Teacher Must Do)
Explain Time required: One to two class periods This will allow students to practice making bar graphs in preparation for creating their climatogram. You will need: Marked plastic cups or graduated cylinders, blue food coloring, water, recording sheet for data, and monthly temperature chart For younger students: Provide students with a prepared climatogram and have them use blocks to represent the amounts that are prefilled on the climatogram for each month.	1. Model the procedure on how to use the clear plastic cup with measurement markings or graduated cylinder while explaining how this will be used to model the collection of the rainfall. 2. Divide students into small groups (preferably their Two Pairs in a Quad) and provide each group with the empty plastic cup with measurement markings or a graduated cylinder, blue food coloring, a cup for pouring the water, and a monthly temperature chart that they will need to place on their climatogram. 3. Demonstrate how to simulate rain by adding a predetermined amount of water (with blue food coloring) to the graduated cylinder or marked plastic cup. 4. Have students practice measuring rainfall in their graduated cylinder or marked plastic cup and recording the measurements on a shared climatogram chart. 5. Ask students to create a bar graph on the climatogram to represent the water data collected over a predetermined time and place the temperature above each graph and connect using lines (see figure 5.12, page 151).
Scaffolds for Fatou: Fatou will be able to demonstrate her understanding by performing the activity. Be sure that, as you model, you are explaining in simple sentences what you are doing and why. Give think time so Fatou can process the steps and information. You can also provide visual instructions for her to follow.	
Elaborate Time required: Two to three class periods You will need: Graph paper for students (have a completed example as a visual) For younger students: The elaborate stage is at the teacher's discretion for grade levels below grade 3. Students in lower grades can show their bar graphs or just their stacking cubes on the prepared precipitation chart.	1. Have groups share their climatograms with the class, explaining the variation in precipitation and temperature over the allotted time provided. 2. Review key vocabulary words and concepts covered in the lesson. 3. Ask students to write a sentence or two about what they learned regarding precipitation and data representation and how geographic features impact weather and climate. 4. Provide a word bank so students can use the vocabulary words.
Scaffolds for Fatou: Sentence frames or starters can be used to help Fatou understand and engage with the material. You can also use visuals for the word bank and vocabulary words. Simplify the language on the climatogram or translate it for her. If Fatou is not comfortable speaking in front of the class, allow Fatou to either present in her small group or just to you, if possible. Ask Fatou yes-or-no questions or have her point to answer your questions about the graph to check for understanding. Since her language proficiency is at the speech emergence level, you can ask her questions at a higher level than yes or no or pointing.	

5E Lesson Plan	Activity (What Teacher Must Do)
Evaluate Time required: One to two class periods You will need: Sentence frames or sentence starters For younger students: Depending on grade level, use sentence starters, ask questions, or use visuals for them to discuss.	1. Formatively assess students' understanding through observation during group activities and discussions. 2. Assess the graphs and sentences produced by students for understanding and application of the lesson's objectives.
Scaffolds for Fatou: Since the evaluation is being done using discussions and activities, allow translation as needed and look at Fatou's work to assess her use of the sentence starters or sentence frames.
Closure Time required: Twenty to twenty-five minutes Connection between the STEM minilesson and the STEM challenge. You will need: Visuals or videos for weather patterns, seasonal changes (keeping in mind some students may come from countries with only one season), and pictures of the water cycle For younger students: Do not connect to water cycle for grades K–2 since it is not taught until grades 3 or 4.	Engage the class in a discussion about the importance of understanding and monitoring precipitation, linking it to broader topics such as weather patterns, and comparing the precipitation averages and seasonal patterns for several locations. Introduce the water cycle to connect the vocabulary words from the STEM minilesson to the STEM challenge. Provide students with a checklist of what should be the takeaways from the STEM challenge, including vocabulary. Have them rate themselves using the checklist as an exit ticket. For example, if there are ten takeaways students should have from the challenge, provide them with a rubric for scoring. If the class takeaway was less than 70 percent then you will need to see where the area of difficulty existed and revisit it the next day to address the gap. This is not for grading purposes, but as a check for learning for you.
colspan="2" Scaffolds for Fatou: Provide visuals for Fatou as you connect the water cycle to vocabulary words. If possible, have a side-by-side comparison of the minilesson and the STEM challenge as you make connections. Stop the video at important parts to explain and allow think time. Allow Fatou to use visuals to interact and to write in Juba Arabic and use a translation app to translate. If Fatou is reluctant to speak in front of the whole class, hold the closure in quads so that she can participate.

FIGURE 5.9: 5E STEM challenge lesson plan.

As seen in figure 5.9, using scaffolds for Fatou within the 5Es instructional model is highly effective because it provides her with structured support at each phase of learning. In the engage phase, scaffolds like visuals and gestures help Fatou connect new content to their prior knowledge. During the explore phase, hands-on activities are paired with sentence frames and bilingual resources, allowing her to actively participate and understand concepts despite language barriers. In the explain phase, graphic organizers and collaborative discussions support Fatou in articulating her thoughts and grasping new vocabulary. The elaborate phase benefits from scaffolds, such as guided practice and modeling, helping Fatou apply her knowledge in new contexts. Finally, in the evaluate phase, scaffolds like checklists and rubrics provide clear criteria for self-assessment, ensuring Fatou can effectively demonstrate her understanding. These scaffolds can also assist teacher reflection and feedback to address gaps in instruction. This strategic use of scaffolding throughout the 5Es enhances comprehension, engagement, and language development for multilingual learners like Fatou.

Figure 5.10 provides a chart students can use to write about a geographic feature and its effect on weather and climate. For Fatou, you can provide pictures for her to choose from, or you can provide the picture and sentence frame and have her state what the picture represents and fill in the sentence frame using words from the word bank.

Geographical Feature	Its Effect on Weather and Climate
1. Mountains 	Possible effect: Large mountain ranges can influence precipitation patterns. Scaffolded for Fatou: _____ mountain _____ can influence _____ patterns. Word Bank: ranges, large, precipitation *Bold words are vocabulary words. **Be sure to use capital letters when beginning a sentence.
2. Oceans 	Possible effect: Places closest to the oceans, particularly those near warm ocean currents, tend to have wetter climates than inland places. Scaffolded for Fatou: Places _____ to oceans or bodies of _____ tend to have _____ than places _____. (Notice that I made the sentences simpler but still maintained the same information.) Word Bank: inland, closer, wetter, water, climates *Bold words are vocabulary words. **Be sure to use capital letters when beginning a sentence.

FIGURE 5.10: Student chart for listing geographical features and the effect on weather and climate.

Figure 5.11 (page 150) contains a Venn diagram for students to complete stating the similarities and differences between weather and climate.

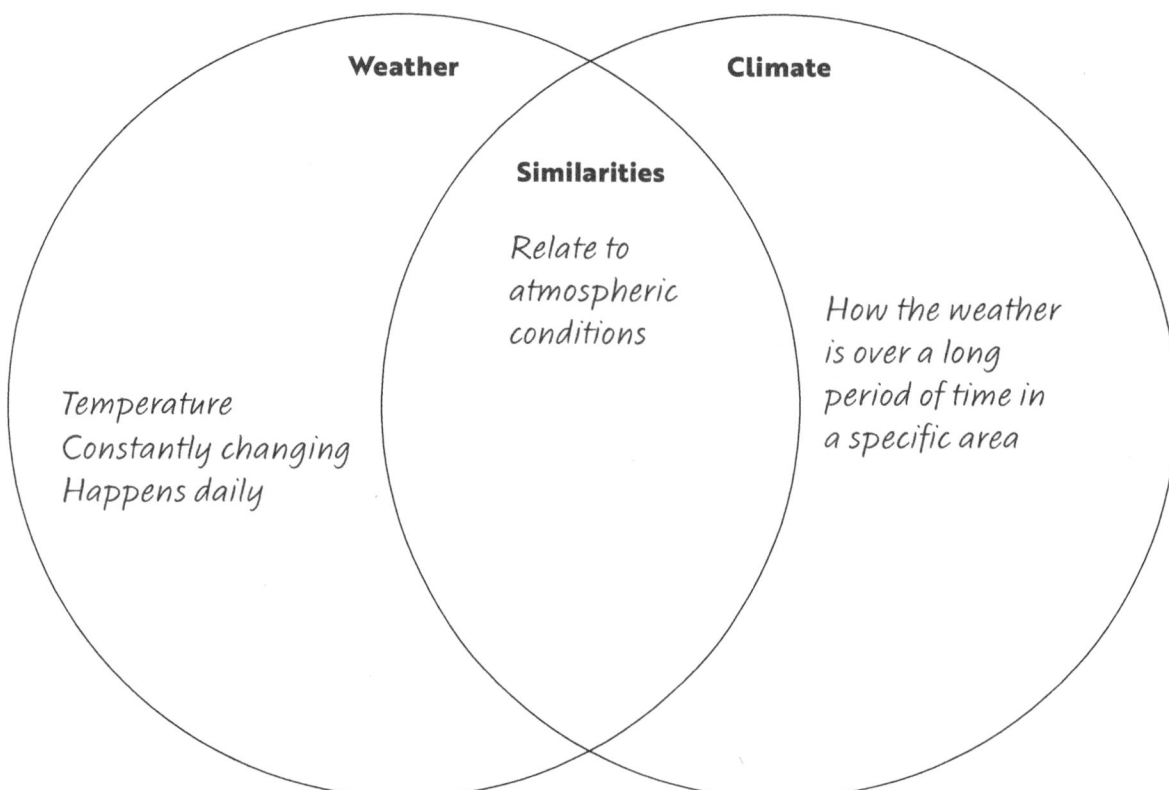

FIGURE 5.11: Venn diagram for writing the similarities and differences between weather and climate.

You can scaffold this task for Fatou by providing her sentence frames to complete or images to place in the diagram. She can use either in discussions with her quad group.

Figure 5.12 contains a sample of a climatogram that the students in grades 4–12 would create for a fictional city using real-world data from Atlantic City (U.S. Climate Data, 2024).

Month	Mean Temperature (°C)	Monthly Precipitation (mm)
January	6	82.0
Februrary	7	73.0
March	11	107.0
April	17	92.0
May	22	85.0
June	27	79.0
July	29	94.0
August	29	104.0
September	25	80.0
October	19	87.0
November	13	83.0
December	8	94.0
Mean	17.8	88.3
Range	23	34

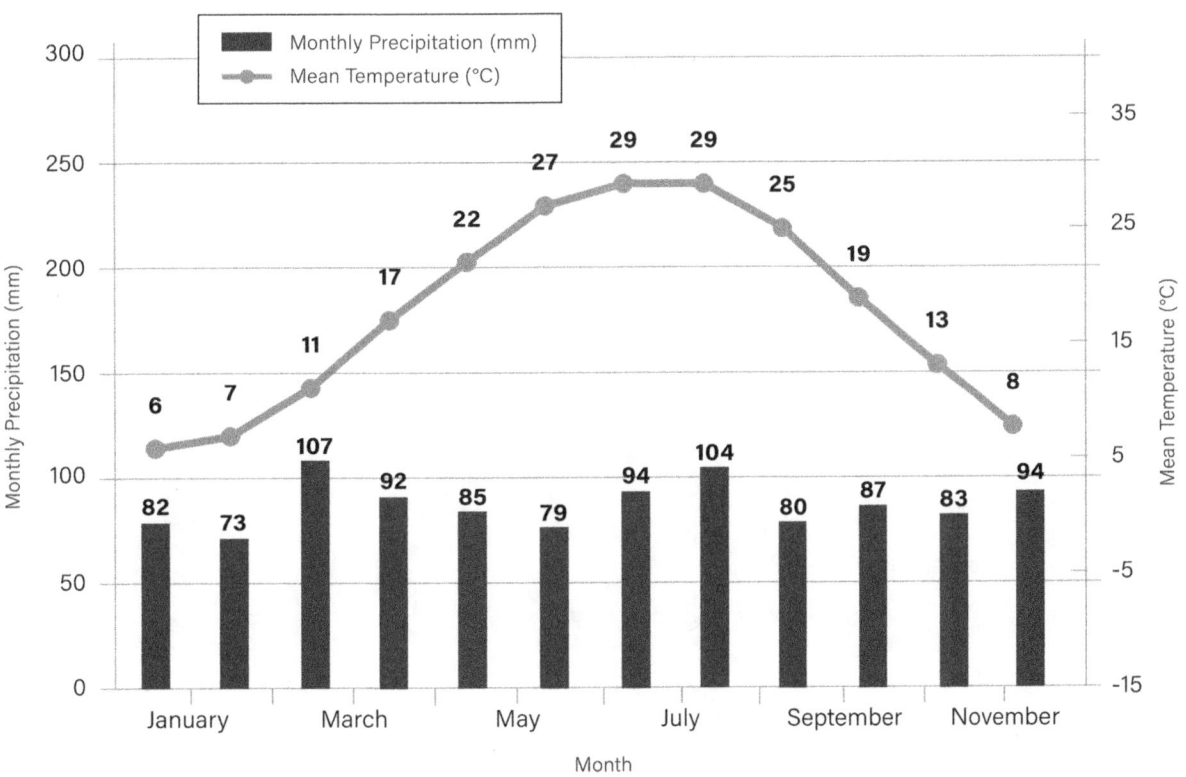

FIGURE 5.12: Sample student climatogram.

Explaining a climatogram (or climograph) to multilingual learners can be challenging, but using simple language, visuals, and examples can make the concept clearer. Here's a step-by-step explanation tailored for multilingual learners.

1. **Introduction with familiar concepts:** Open the topic with common knowledge. "You know how weather changes, right? Some days it's hot and some days it's cold. Sometimes it rains and sometimes it doesn't."

2. **Visual aid:** Show a sample climatogram. "This is a climatogram. It's a chart or graph that shows the average temperature and rainfall for each month in a specific place."

3. **Temperature explanation:** Point to the temperature curve. "This line tells us how hot or cold it is each month. It works like a thermometer. The higher the line, the hotter it is. The lower the line, the colder it is."

4. **Rainfall explanation:** Point to the rainfall bars. "These bars tell us how much rain falls each month. Like a bucket collecting rain. Taller bars mean more rain. Shorter bars mean less rain."

5. **Compare and contrast:** "By looking at this, we can see which months are hot and rainy or cold and dry. For example . . ." (show a specific month.)

6. **Purpose and usage:** Explain how the graph can be used. "People use climatograms to understand the climate of a place. If you're traveling or farming, it's good to know when it's going to be hot or if there will be rain."

7. **Interactive assessment:** Get your students involved. "Can you point to the hottest month? How about the month with the most rain?"

   By breaking down the explanation step by step, utilizing visuals, and engaging learners interactively, it becomes easier for a multilingual learner to grasp the concept of a climatogram.

Select climatograms from various cities with different geographical features. Have students work in their Two Pairs and Quad to complete the paragraph, like the one shown in figure 5.13.

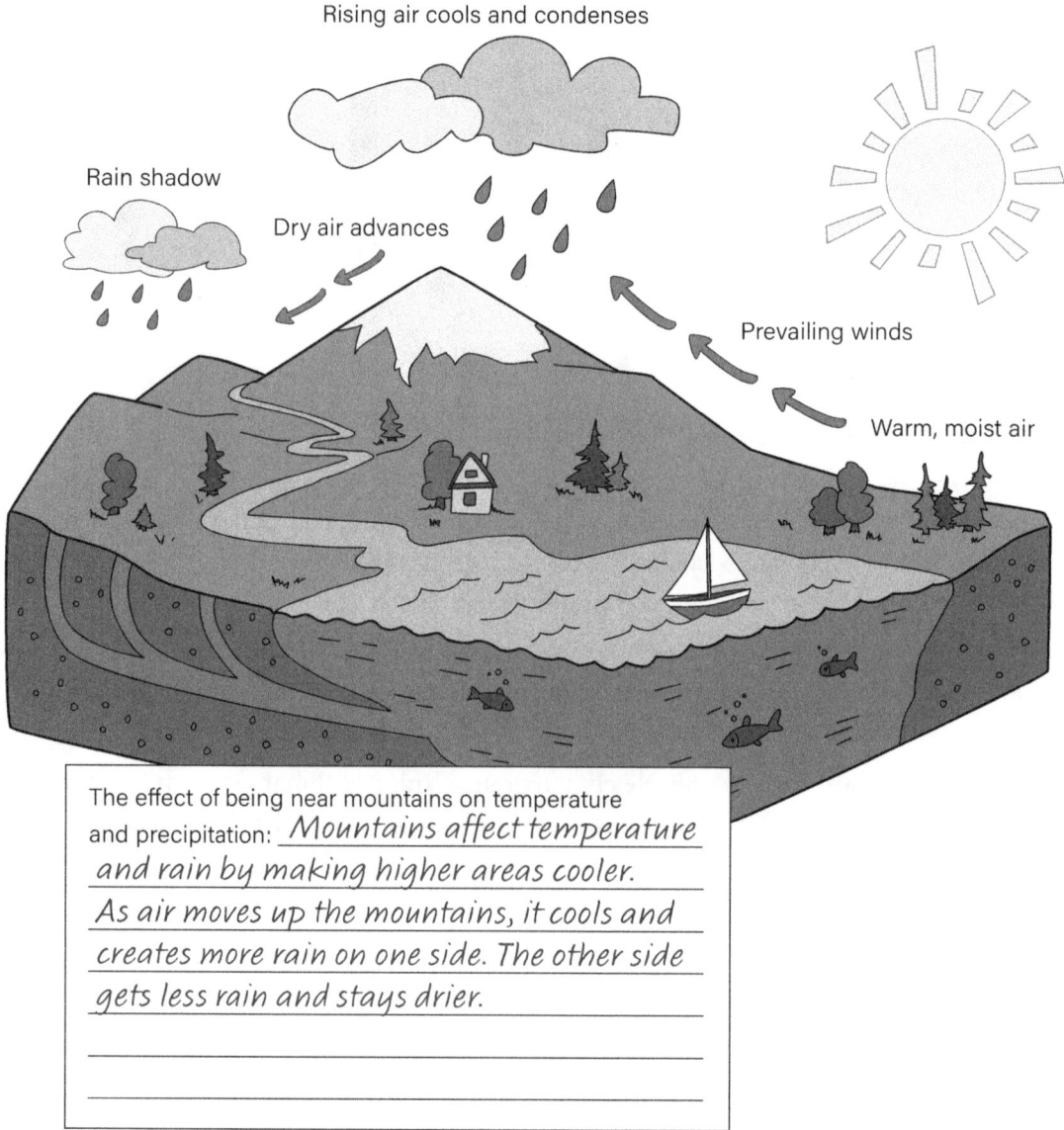

FIGURE 5.13: Sample image and sentence starter.

Students continue to work in their quad group to complete the STEM challenge. You will notice the format and routine of the STEM minilesson is the same for the STEM challenge. Providing familiar formats and routines to lessons offers predictability for multilingual learners. A stable environment helps them to feel comfortable because they know how things will be. They know what comes next, and they will not have to worry about foundational things. This is important, because if students don't feel comfortable, they are not going to take risks or make mistakes, which reduces their potential to learn and grow (Colorin Colorado, 2018). By providing formats and routines, you give multilingual learners a road map for important moments during the lesson and allow them to internalize and take ownership of their choices and move quickly into new learning experiences. This internalization, engagement, and ownership is achieved through the mindful scaffolding of routines (EL Education, n.d.).

Key Takeaways

Leveraging multilingual learners' assets is essential because it recognizes and values their unique linguistic, cultural, and experiential contributions, enriching the learning environment for all students. By building on multilingual learners' strengths, educators can create more inclusive and responsive instructional strategies that cater to diverse learning styles and perspectives. Utilizing minilessons not only enhances multilingual learners' academic performance but also boosts their confidence and sense of belonging by preparing them for the main STEM challenge and equipping them to be actively involved.

Consider the following key takeaways from this chapter.

- Using the STEM minilesson prior to the actual STEM challenge offers the teacher insight into the multilingual learner's background knowledge and gives the learner background knowledge in the event they don't have experience with the content.

- Storytelling is a powerful tool in teaching multilingual learners because it engages students emotionally and cognitively, making learning more memorable and meaningful. Stories provide context and make abstract concepts concrete, helping multilingual learners understand and retain new vocabulary and language structures.

- By tapping into students' funds of knowledge, educators can create a more inclusive, engaging, and effective learning environment that honors and utilizes the rich resources each student brings to the classroom.

- Effectively activating and assessing multilingual learners' prior knowledge ensures that instruction is both accessible and engaging and promotes a deeper connection to new learning content.

- The 5E model for teaching multilingual learners provides a structured approach to learning that includes engaging students with relevant content, allowing them to explore through hands-on activities, and explaining concepts with clear, accessible language. It further supports learning by elaborating through practical applications and evaluating understanding with ongoing assessments that ensure comprehensive language development and content mastery.

CHAPTER 6

Using Claim, Evidence, and Reasoning to Build Language Fluency

> *One of the best ways for English Learners to improve their language development is through the use of the language.*
>
> **—NATIONAL ACADEMY OF SCIENCES**

All students are curious no matter where they are born or what language they speak. Curiosity—the desire to explain how the world works—drives the questions we ask and the investigations we conduct. But how do we plan lessons to allow students' natural curiosity to run wild in the science classroom when their first language is not English? And how do we create an environment where all students can interact like scientists where language and content are taught simultaneously?

One answer is by designing inquiry-based science instruction using the claim, evidence, reasoning (CER) model. CER is one way to plan engaging lessons around a student's natural curiosity while keeping them engaged. However, given that the CER format of writing explanations can be challenging even for native English speakers, imagine how much more challenging that might be for multilingual learners. To successfully implement CER, teachers must explicitly introduce and model it for students.

In this chapter, I provide an overview of CER's four-part framework and explain why it is a powerful tool to support multilingual learners in building language proficiency. I also show how to introduce this model to your classroom because students require consistent practice to become familiar with using the framework. Using two examples, I will show you what CER looks like in action.

CER's Four-Part Framework

CER was developed for science education by the National Research Council (McNeil & Luft, 2021). It is a structured form of scientific argumentation (one of eight scientific practices in the Next Generation Science Standards) that, when used intentionally and purposefully, can provide the practice multilingual learners need toward simultaneously building proficiency and content. Scientists' most fundamental skills are asking questions, making predictions, designing investigations, analyzing data, supporting claims with evidence, and debating conclusions with other scientists. The CER model is a useful tool because constructing an explanation and forming a conclusion based on evidence are essential skills in both science and engineering. The use of argumentation in science, which is an important part of scientific inquiry and underpins reasoning across STEM domains (Erduran, Ozdem, & Park, 2015), is also a collaborative learning exercise rooted in a cognitive process (Salter & Renken, 2017). Using CER in science instruction provides multilingual learners with multiple modalities and registers for practicing English. Using CER allows all students to work like scientists while multilingual learners can simultaneously learn science-based English in its proper context.

CER is quite similar, in my opinion, to the structure and purpose of three-act mathematics tasks. Three-act mathematics tasks were introduced to the world in 2010 by high school teacher Dan Meyer (Fletcher, 2016). These tasks are designed to get students thinking critically, collaborating, and working cooperatively (see chapter 4, page 78, for more about cooperative learning). Three-act mathematics tasks are a whole-group mathematical task that consists of three distinct parts or acts (see figure 6.1).

- **Act 1:** Provides an engaging or perplexing visual
- **Act 2:** Provides additional information to assist in seeking a solution
- **Act 3:** Provides the solution or what is called "the reveal"

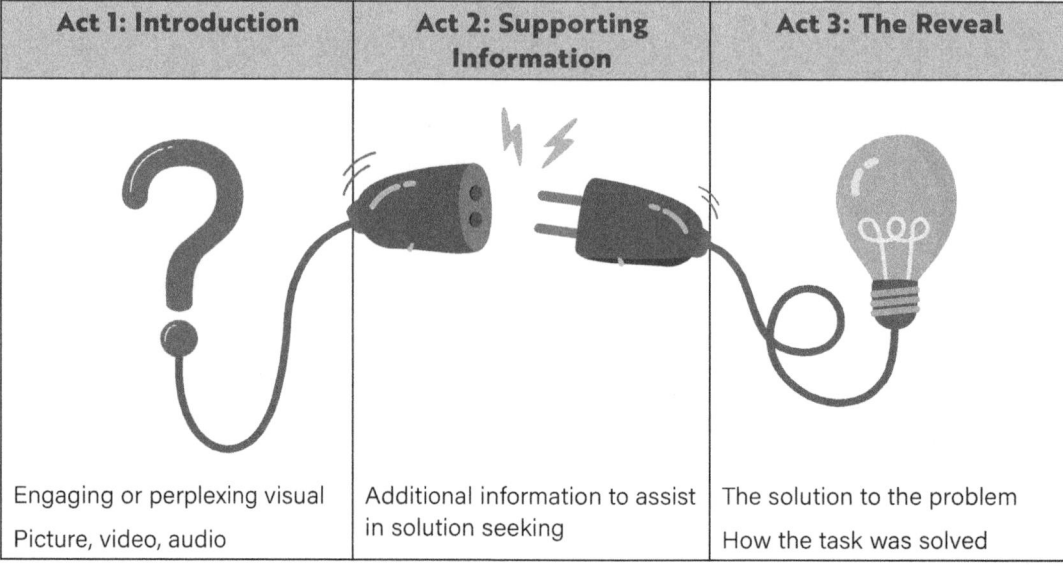

FIGURE 6.1: Three-act mathematics task structure.

The CER model consists of four distinct parts as opposed to three distinct parts in three-act mathematics tasks. The four parts of CER are as follows.

- **Part 1: Question and vocabulary**—The teacher asks a close-ended question after providing a visual that can be either video, picture, and so on, based on phenomena or that connects with a lab assignment. Vocabulary is introduced during this part as well, but not explicitly taught. The students' answers to the question initiates the three-part structure of CER: the claim, the evidence, and the reasoning.

- **Part 2: Claim**—The students state their opinion or observation to the question based on the visual the teacher provided. It will only be one sentence. When stating a claim, there is no explanation, reasoning, or evidence.

- **Part 3: Evidence**—This is the data used to support the claim. The students collect the data they need to answer the question from the claim. They need to decide how they can collect the data and how they can investigate it. It can be quantitative or qualitative. Students should have, at a minimum, two pieces of evidence, but three pieces of evidence creates a pattern. It is important for the teacher to stress to students that the data they collect may disagree with their original claim and that is fine. The purpose of collecting evidence is to provide justifications for their claim.

- **Part 4: Reasoning**—It is based on the evidence students collect once they have conducted the investigation and have data (this is why it is between claim and evidence in the following figure). The explanation is based on a rule, scientific principle, law, or definition that describes why the evidence supports the claim. In other words, how *logically* does the evidence support the claim.

Figure 6.2 illustrates the relationship between CER's four parts.

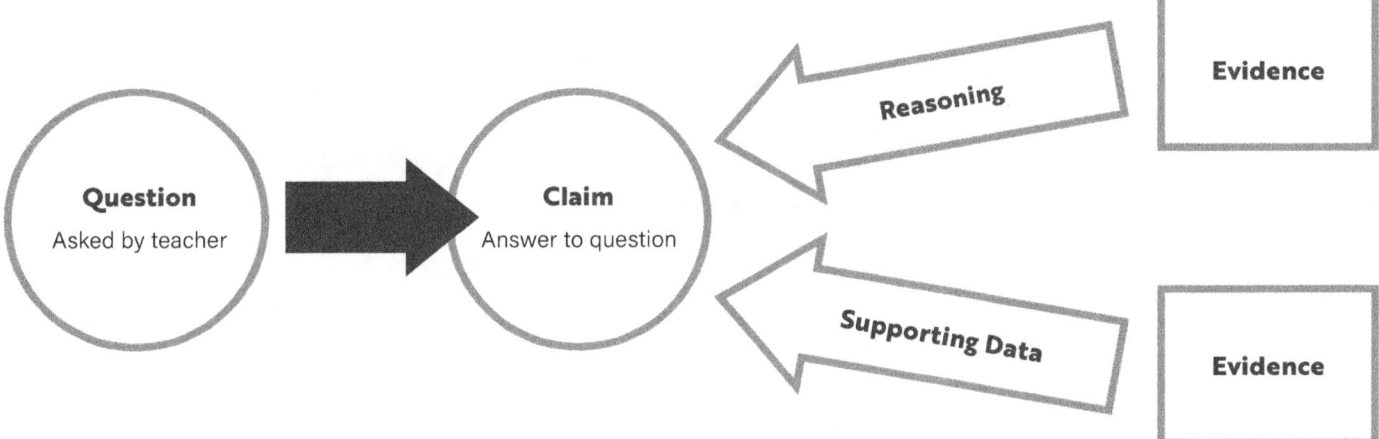

FIGURE 6.2: The four parts of claim, evidence, and reasoning (CER).

Table 6.1 (page 158) compares three-act mathematics tasks and the claim, evidence, reasoning model.

TABLE 6.1: Comparison Between Three-Act Mathematics Tasks and CER Model

	Three-Act Mathematics Tasks	Claim, Evidence, Reasoning Model
Part 1 or Act 1	The teacher provides an engaging or perplexing visual.	The teacher asks a close-ended question based on phenomena or connects with a lab assignment
Part 2 or Act 2	The teacher provides additional information to assist in seeking solutions.	Claim: The students state their opinion or an observation to the question that is only one sentence and does not contain an explanation, evidence, nor reasoning.
Part 3 or Act 3	The teacher provides the solution or what they call "the reveal."	Evidence: Students provide data that can be quantitative or qualitative. Should have a minimum of two pieces of evidence, but three pieces of evidence is best.
Part 4 CER only		Reasoning: The students provide an explanation about how logically their evidence supports the claim.

Three-act mathematics tasks and the CER model both emphasize inquiry-based learning and critical thinking. In a three-act mathematics task, students begin with a real-world problem (Act 1) that ignites curiosity and questions. Next, they gather and analyze data (Act 2) to explore potential solutions. Finally, they create arguments and present their findings (Act 3), similar to making a claim, backing it with evidence, and explaining their reasoning. This method parallels the CER model, where students make a claim, use gathered evidence, and explain their reasoning to connect the evidence to the claim. Both approaches foster deep thinking, justification of conclusions, and a solid understanding of mathematicsematical concepts through structured exploration and reflection. Both methodologies support multilingual learners by encouraging active participation, collaborative learning, and the use of language in context. This helps them build language skills alongside their understanding of mathematicsematical concepts through structured exploration and reflection as discussed in the following section.

How CER Helps Multilingual Learners Build Language Proficiency

Multilingual learners and their native English-speaking peers face the same challenges in the science classroom. They must both use language like a scientist and collaborate to support growth in scientific understanding (Miller, Lauffer, & Messina, 2014). But multilingual learners need to learn to do this in a language that is not their native language. The CER process allows multilingual learners opportunities to speak, listen, read, and write—the four fundamentals in building language proficiency—to engage in meaningful discourse about science or any of the STEM content classes. CER provides students with authentic language opportunities instead of focusing on isolated facts and vocabulary. CER engages students by providing language-learning opportunities because they are learning how to talk about their ideas and share their discoveries with peers while still learning the content and the language.

Sentence starters assist multilingual learners in sharing their ideas and reasonings with their peers by providing opportunities for collaborative learning and language practice. Consider the following examples as a way sentence starters promote meaningful communication and interaction in the classroom.

- I agree with you because _____.
- I don't agree with you because _____.
- What is your evidence that _____?

Sentence starters are scaffolds that provide structure and support for students in developing their ability to analyze and write as they progress academically. Offering sentence starters also decreases anxiety for all students (not just multilingual learners) because it shifts the focus from language and writing to learning the content. Since CER is challenging for native speakers as well, sentence starters are beneficial for all students because they ensure that everyone is part of the learning community. AI programs, such as ChatGPT and Magic School, can create sentence starters for you. All you need to do is input the topic or standard and ask it to create sentence starters or sentence frames for you.

To use Magic School, select Sentence Starters from the menu for grade level and input "Claim, Evidence, and Reasoning" in the input window. I received the following output when I did this for third grade (Magic School AI, n.d.).

1. My claim is _____ because . . .
2. I know this is true because the evidence shows . . .
3. This supports my claim because . . .

To use ChatGPT, ask for sentence starters for multilingual learners by grade level. I received the outputs displayed in figure 6.3 when I did this for third grade.

Claim	Evidence	Reasoning
I think that . . .	I found that . . .	This means that . . .
I believe that . . .	My evidence is . . .	This shows that . . .
My claim is . . .	I observed that . . .	This proves that . . .
In my opinion . . .	According to the data . . .	This is because . . .
It seems to me that . . .	One example is . . .	

Source: OpenAI, 2024.

FIGURE 6.3: ChatGPT-generated sentence starters.

These simplified sentence starters can help multilingual learners express their thoughts and arguments more comfortably.

CER in Your Classroom

When introducing the CER model to the class, remember that students will need multiple opportunities to practice, a lot of support and guidance, and time to learn how to write a great CER. Creating CERs is a time-consuming task, but a worthy and valuable skill for developing critical thinking and scientific inquiry. Teaching students the CER model supports their scientific-inquiry skills and contributes to the development of broader cognitive and communication skills that are essential for academic and professional success.

Explicit modeling and explanations will be needed consistently throughout the CER learning process in addition to effective strategies, such as modeling, repeating, asking for clarification, rephrasing, and building on what students say. These strategies will enable multilingual learners to better understand the process.

One way I recommend introducing the CER process to your multilingual learners is to first use a nonscience topic (see "CER in Action" on page 163), like images, artwork, videos, a popular TV commercial, or even an introduction to a popular new video game. Using nonscience topics will relieve a lot of anxiety for students because it takes the focus off science and language content and places it fully on the CER process. The first time you do this, have students work in small or paired collaborative groups (such as the cooperative learning groups discussed in chapter 4, page 93) as you explicitly model the entire process with the class. You may need to do this multiple times over the course of a few days to ensure that students understand the process and expectation. Remember, learning the CER process takes time to master, so be patient but remain diligent. The skills it provides for students makes it worth the time.

Figure 6.4 shows a CER template you can use. You can also create a different one for your students. The sample template includes possible sentence starters for each part of the process.

	Sentence Starters	Student's Written Reply
Claim	I believe that . . . In my opinion . . . From what I know . . . I think this because . . . It seems to me that . . .	
Evidence 1	For instance . . . An example is . . . I can show this with . . . According to the text . . . This is supported by . . .	
Reasoning 1	This matters because . . . The reason for this is . . . If you think about it . . . To explain further . . . In simple terms . . .	

Evidence 2	For instance . . . An example is . . . I can show this with . . . According to the text . . . This is supported by . . .	
Reasoning 2	This matters because . . . The reason for this is . . . If you think about it . . . To explain further . . . In simple terms . . .	

FIGURE 6.4: Claim, evidence, and reasoning student template.

*Visit **go.SolutionTree.com/EL** for a free reproducible version of this figure.*

Figure 6.5 includes a CER rubric you can share with students to communicate performance expectations for each phase of the CER process.

Claim, Evidence, Reasoning Rubric

Note: This rubric assesses students' ability to create a claim supported by evidence and reasoning. The rubric uses a three-point scale to evaluate the clarity and specificity of the claim, the relevance and accuracy of the evidence provided, and the strength of the reasoning connecting the claim and evidence.

Criteria	3 points	2 points	1 point	0 Points
Claim	The claim is clear, specific, and well-supported by evidence and reasoning.	The claim is somewhat clear, specific, and supported by some evidence and reasoning.	The claim is unclear, vague, or unsupported by evidence and reasoning.	Nothing turned in or it was blank.
Evidence	The evidence provided is relevant, accurate, and effectively supports the claim.	The evidence provided is somewhat relevant, accurate, and supports the claim to some extent.	The evidence provided is irrelevant, inaccurate, or does not effectively support the claim.	
Reasoning	The reasoning provided demonstrates logical thinking, analysis, and a strong connection between the claim and evidence.	The reasoning provided demonstrates some logical thinking, analysis, and a partial connection between the claim and evidence.	The reasoning provided is illogical, weak, or does not effectively connect the claim and evidence.	

FIGURE 6.5: CER rubric.

*Visit **go.SolutionTree.com/EL** for a free reproducible version of this figure.*

Figure 6.6 contains a CER checklist you can use to evaluate student performance or simply remind students what should be present in each phase of the process. Feel free to customize this checklist based on your specific requirements or add any additional criteria that you find important.

CER Checklist	Yes or No
Claim	
Is the claim clear and specific?	
Does the claim relate to the topic or prompt?	
Is the claim supported by evidence and reasoning?	
Evidence	
Is the evidence relevant to the claim?	
Is the evidence accurate and reliable?	
Does the evidence effectively support the claim?	
Reasoning	
Does the reasoning demonstrate logical thinking?	
Is there a clear connection between the claim and evidence?	
Does the reasoning provide analysis or explanation?	

FIGURE 6.6: CER checklist.

*Visit **go.SolutionTree.com/EL** for a free reproducible version of this figure.*

Remember to always allow students, especially multilingual learners, to collaborate in small groups or pairs, such as face partners, shoulder partners, and quads discussed in chapter 4 (page 84). Collaborative grouping provides students the opportunity to hear and practice academic language in context. The collaborative groups should be a heterogeneous grouping of student language abilities so multilingual learners can build academic vocabulary and practice the language in a safe small-group setting.

Continue using a template and allowing students to work in pairs or small groups as they progress in using the CER model. Consider the following ideas for assessing student understanding.

- **Beginning CER skill level:** The teacher only provides the question and students must make a claim and provide the evidence. As a class, write the reasoning together with everyone sharing.

- **Intermediate CER skill level:** The teacher tells students who are working in assigned collaborative groups, to write the evidence and reasoning based on a question and claim the teacher created.

- **Advanced CER skill level:** The teacher will have students complete the entire process, including creating the question, when the teacher feels they are ready. This will allow the teacher to determine whether the students completely understand how to use the CER process.

Pause for a moment and reflect on what you've read so far. Review figure 6.1 (page 156). Why do you think reasoning is located between claim and evidence when evidence is the last phase of the CER model?

CER in Action

Now that you know how to introduce CER in your classroom, let's imagine what it might look like to use the CER template. We'll look at two examples in this section. The first example shows a sample completed CER template showing how four students at different levels of language proficiency (beginner, intermediate, and advanced/native) can respond.

Example 1: Gorilla® Packaging Tape TV Commercial

In the first example, the nonscience video is a TV commercial for Gorilla packaging tape. The template is broken up into four sections: (1) question, (2) claim, (3) evidence, and (4) reasoning. Notice that evidence has sections for two pieces of evidence. There may be as many as three evidence statements or as few as one statement. It is your template, so you can decide. You may want to start with one evidence statement for simplicity, and as the students become more comfortable with the process, move to two or even three evidence statements if you want students to find more data to support their claim. You can easily tailor the template to your class's needs.

For this practice exercise, the link to the thirty-second commercial is provided for you to view. After watching the commercial, review the completed CER template (figure 6.7) and think of how your students will respond, the various levels of language abilities in your class, and what sentence frames you should provide.

Example 1: TV commercial—Gorilla packaging tape.

Question: Is Gorilla packaging tape the best for the toughest jobs on planet Earth?™

Gorilla Packaging Tape Commercial
www.youtube.com/watch?v=Hz12277AynI

CER	Sentence Starter	Student Statement
Claim: Student responses should be more than a yes-or-no answer, except for beginner English learners. All other students should use complete sentences or short responses.	I believe that _____. It is my opinion that _____. I observed _____ when _____.	**Beginner English learner:** "No." **Intermediate English learner:** "I believe that Gorilla tape is not the toughest on Earth." **Advanced English learner or native speaker:** "I do not believe that Gorilla tape is the best for the toughest jobs on planet Earth."

Note: Because this is a commercial, no experiment is going to be conducted. Only the visuals from the video can be considered as their data.		
Evidence 1: What specific observation or data did you use to support your claim? There can be multiple sections for evidence. If so, the students must also have reasoning to support each piece of evidence.	The commercial states _____ . The evidence from the commercial _____ . According to the commercial _____ . The commercial uses _____ and not a _____ .	**Beginner English learner:** "The commercial uses a box and not a heavy box." **Intermediate English learner:** "The evidence from the commercial only used a box of books and papers, not something heavy." **Advanced English learner or native speaker:** "The commercial did not use anything that was a tough job. Taping a large box full of notebooks and paper is not what I consider a tough job."
Evidence 2: What specific observation or data did you use to support your claim?	The commercial states _____ . The evidence from the commercial states that _____ . According to the commercial _____ .	**Beginner English learner:** Draws picture of a metal box, and a box that is not heavy. Labels the parts and uses the words *heavy* and *light* from the word bank the teacher provided. **Intermediate English learner:** "The commercial states the width and thickness of the tape but does not say how this makes it better." **Advanced English learner:** or native speaker: "Just because tape is wider and thicker does not mean it can do jobs that are bigger or heavier. You cannot assume that wider means better."
Reasoning: How does your reasoning support your claim? What science principle links your reasoning to the claim? Minimum of two pieces of evidence.	If _____ , then _____ . This proves _____ because _____ . This supports my claim because _____ . This confirms that _____ because _____ .	**Beginner English learner:** Uses the pictures drawn but shows the tape break or the box of items fall out of the box. "This supports my claim because box breaks." (Grammatical errors are common at this stage.) **Intermediate English learner:** "This supports my claim because the width and thickness of the tape does not mean it can do the toughest jobs. There was no evidence in the video to support their statement. This confirms my claim that the Gorilla tape is not toughest on Earth because they only used a box and nothing to else to support the question." **Advanced English learner or native speaker:** "The reasoning that supports my claim that Gorilla tape is not the best for the toughest jobs on planet Earth is because the commercial only used a box with light items. Evidence was not provided to show how the tape would perform on different materials or weights. There was no evidence that connected the width and thickness made a difference on its ability to do tough jobs."

FIGURE 6.7: Completed commercial CER practice—Gorilla packaging tape.

Think about how students will use the sentence starters in their responses, and how their reading, writing, speaking, and listening abilities align with what the student can do at their level of language proficiency. Tables 6.2 (page 165), 6.3 (page 166), 6.4 (page 166), and 6.5 (page 167) explain what multilingual learners can do at various stages of language development for students in grades 2 and 3. These grades were selected because the next student I'll introduce, Wang Xiu Ying, is in grade 3—using skills they will need when completing a claim, evidence, and reasoning activity. The skills are retelling, explaining, arguing, and discussing. It is important to understand these differences in abilities so that your expectations match their ability. Remember, language ability does not define academic ability. This is not an exhaustive list. Please see WIDA Can Do Descriptors (WIDA, n.d.a) for a full list.

TABLE 6.2: Recounting

Recounting: To retell to display knowledge or narrate experiences or events (claim)			
Entering	**Developing**	**Expanding**	**Reaching**
Students can:	**Students can:**	**Students can:**	**Students can:**
Reading: Identify key phrases and words in illustrated text	**Reading:** Illustrate experiences of characters in illustrated statements and identify time-related language in context	**Reading:** Put a series of events in order based on familiar texts, identify main idea in illustrated text, paraphrase with support, and highlight relevant information	**Reading:** Identify setting and character from grade-level text and determine central message from fables and folktales
Writing: Label illustrations that illustrate steps and create visual representations	**Writing:** List ideas using graphic organizers and describe visual information	**Writing:** Describe a series of events, create stories with details about characters, and provide details and examples about narratives	**Writing:** Signal order of events using temporal words and phrases and relate real or imagined experiences or events
Listening: Show what happens next by pointing or drawing and create visual displays based on oral prompts	**Listening:** Identify who, what, where, and when of illustrated statements and main materials from oral descriptors	**Listening:** Reenact and identify content-related situations from oral descriptions and multimedia, retrace steps of a process, and make designs or models from oral directions	**Listening:** Identify key ideas from the text read aloud or presented orally and determine main idea of texts read aloud in diverse media and formats
Speaking: Respond to questions related to stories and act out or name events throughout a school day	**Speaking:** Reproduce facts in context and participate in multimedia presentations based on research	**Speaking:** Sequence events in stories, describe situations from school or community, ask and answer questions from speakers, and describe main idea	**Speaking:** Provide descriptive details in content-related activities and name the steps for producing multimedia presentations with some detail

Source: © 2016 by WIDA. Used with permission.

TABLE 6.3: Explain

| \multicolumn{4}{l}{Explain: To clarify the why or how of ideas, actions, or phenomena (evidence)} |

Entering	Developing	Expanding	Reaching
Students can:	**Students can:**	**Students can:**	**Students can:**
Reading: Identify words or phrases and match pictures with graphics in titles or highlighted texts	**Reading:** Interpret images, illustrations, and graphics; sequence sentences; and locate detail in content	**Reading:** Illustrate cause-and-effect relationships in content area and identify relevant information from texts on the same content-area topic	**Reading:** Describe the connection between a series of historical events, scientific ideas, or steps in technical procedures in texts
Writing: List or illustrate ideas, and state facts associated with images or illustrations	**Writing:** Describe elements of processes and procedures, state how something happens using illustrations, compare causes of different phenomena, and state ideas about content-related topics	**Writing:** Describe strategies to solve problems and details of processes, procedures, and events	**Writing:** Elaborate on topics with facts, definitions, and details
Listening: Point to objects from oral clues and pair objects, pictures, or equations directed by a partner	**Listening:** Carry out steps and compete graphic organizers described orally and follow simple sequences presented orally	**Listening:** Identify content-related ideas and details in oral discourse and follow a series of short oral directions to create models of content-area phenomena or processes	**Listening:** Compare strategies from extended oral discourse
Speaking: Describe outcomes with guidance or visuals and answer "Wh-" questions related to classroom routines	**Speaking:** Name steps in process and procedures, describe familiar phenomena in words or phrases, and express cause and effect of behaviors or events	**Speaking:** Describe consequences of behaviors or occurrences and elaborate on the cause of various phenomena	**Speaking:** Synthesize main ideas from supporting details of text read aloud or information obtained from diverse media

Source: © 2016 by WIDA. Used with permission.

TABLE 6.4: Argue

| \multicolumn{4}{l}{Argue: To persuade by making claims supported by evidence (reasoning)} |

Entering	Developing	Expanding	Reaching
Students can:	**Students can:**	**Students can:**	**Students can:**
Reading: Identify facts in illustrated informational text read orally	**Reading:** Distinguish fact from fiction (for example, using sentence strips or highlighting texts), identify claims or opinions in illustrated texts, and identify general academic and content-related words and phrases in text relevant to the genre	**Reading:** Illustrate cause-and-effect relationships in content area and identify relevant information from texts on the same content-area topic	**Reading:** Evaluate characters, settings, and events from a variety of media
Writing: Indicate decisions or preferences through labeled pictures, words, or phrases and provide evidence of natural phenomena or opinions through labeled drawings	**Writing:** Connect preferences, choices, or opinions to reasons and describe pros and cons related to social issues or familiar topics.	**Writing:** Describe strategies to solve problems and details of processes, procedures, and events	**Writing:** Elaborate on opinions and reasons, as well as compare and contrast important points and details presented in two texts on the same topic
Listening: Indicate personal points of view in response to oral phrases or short sentences (for example, by thumbs-up or thumbs-down and agree or disagree cards)	**Listening:** Distinguish opinions from facts from peers' oral presentations and identify different points of view in short oral dialogues	**Listening:** Interpret oral information from different sides, identify opposing sides of arguments in dialogues, and identify claims in oral presentations	**Listening:** Identify evidence to support claims and opinions from multimedia and follow agreed-on rules for discussions around differing opinions
Speaking: State a claim or position from models or examples and share facts as evidence using sentence starters or sentence frames	**Speaking:** Ask and answer questions in collaborative groups, tell what comes next and show why, and share reasons for opinions or claims (for example, science experiments)	**Speaking:** Defend claims or opinions to content-related topics, express and support different ideas with examples, and provide evidence to defend own ideas	**Speaking:** Connect personal comments to others' remarks to build a case for ideas or opinions and summarize ideas or opinions from two sides.

Source: © 2016 by WIDA. Used with permission.

TABLE 6.5: Discuss

Discuss: Interacting with others to build meaning and share knowledge (claim, evidence, and reasoning)			
Entering	**Developing**	**Expanding**	**Reaching**
Students can:	**Students can:**	**Students can:**	**Students can:**
Express own ideas through drawings, gestures, words, and phrases and express agreement or disagreement nonverbally (for example, thumbs-up or thumbs-down)	Ask yes-or-no questions to request clarification, recognize how different intonations convey different meanings, negotiate agreement in small groups, and express own ideas consistent with the topic discussed	Express own ideas and support ideas of others; propose new solutions to resolve conflict in small groups; initiate and maintain conversations; challenge ideas respectfully and listen to, build on, and extend ideas	Share topic-related information, build on others' remarks by linking comments, and maintain audience engagement through specific language and body movement

Source: © 2016 by WIDA. Used with permission.

In the following section, we'll look at a second example, demonstrating one way the teacher could model the CER process to the entire class.

Example 2: Whole-Class Introduction Using Gorilla Packaging Tape Commercial

In the second example, the teacher introduces the CER model to the whole class using a nonscience video. The teacher created a visual of any vocabulary words from the commercial that may be challenging or confusing for the class's multilingual learners and reviewed the vocabulary words and their meanings with the class prior to showing the commercial. We will be using the Gorilla packaging tape commercial again.

Pay particular attention to the following.

- How many times the commercial will be played
- How the teacher makes a connection to explain argumentation, a word that can be problematic for multilingual learners
- How the activity progresses to the main purpose of creating a CER statement
- How explicit the teacher is when describing the instructions for the question, claim, evidence, and reasoning
- How the teacher is modeling their thinking aloud for the students to hear
- How the students interact with each other

Note: The sections in brackets provide additional background information.

INTRODUCE ARGUMENTATION

Let's look at an example of argumentation.

> **Teacher:** *Have any of you ever argued with a brother, sister, or friend? Can anyone give me an example? [Teacher shows a picture of two people arguing.]*
>
> **Joel:** *Yes. I argue with my sister because she is always on the computer playing her video games.*

> ***Teacher:*** *What would you think if I told you that scientists also argue with each other? Well, they do. But not like Joel described about his sister. When scientists argue, it is called, argumentation and we are going to learn about argumentation today. [The teacher writes the word* argumentation *on the board, the students have the word at their desks with a visual, and if possible, have the word translated into the languages spoken in class.] Argumentation in science follows special rules or framework. It is called CER and it stands for claim, evidence, and reasoning. [The teacher writes this on the board.]*
>
> ***Teacher:*** *CER is what we are going to be learning about today and we will use it throughout the school year. CER is how we will learn to agree or disagree with each other respectfully and work like scientists. To better understand how to use CER for argumentation, we are going to watch a short commercial that many of you may have seen before. I will play it three times.*

The teacher places students in small groups of two to four students so that the multilingual learner will be more willing to converse. The teacher explains what the students should do each time the video plays: The first time is just for the students to listen to it. The second time is for students to write down what they notice and wonder. The third time is for students to discuss in their groups what they think the advertisers are trying to get them to believe. The goal is to get students, especially multilingual learners, used to sharing their thoughts in a comfortable environment.

IMPLEMENT THE CER PROCESS

Now that you have been introduced to the process and purpose of argumentation and have received an example of how to introduce the concept, you will learn how to implement the CER process.

> ***Teacher:*** *You will be working in your quad. [Plays the commercial for the first time to the class during the lesson opening.] OK. Now that you have watched the commercial for the first time, I want you to watch it again and share with your group what you notice and what you wonder. I have displayed sentence starters on the board for you to use if needed.*

Figure 6.8 includes sample sentence starters.

After the teacher has played the commercial for the second time and reviewed and modeled how to use the sentence starters for one person to use and for the group, the teacher walks around the classroom listening to the students share what they notice and wonder about. One of the groups the teacher is listening to has Shalleen, a beginner multilingual learner from Mexico; Omar, an intermediate multilingual learner from the Middle East; and David and Zinya, native English speakers.

> ***Shalleen:*** *[Draws a picture of a gorilla.]*
>
> ***Omar:*** *I noticed that the box broke and all the stuff came out.*
>
> ***David:*** *I noticed that the box was too big for the lady to carry by herself.*

> 1. I noticed _____ .
> 2. I noticed in the commercial that _____ .
> 3. I wonder _____ .
> 4. I wonder whether _____ .
> 5. We wondered _____ .
> 6. We noticed that _____ because _____ .

FIGURE 6.8: Sample sentence starters.

> *Zinya:* I noticed that the box looked very thin and flimsy.
>
> *Teacher:* So, Shalleen, you noticed a gorilla. Now discuss as a group what you wonder. When you're done, prepare a notice-and-wonder statement for the group to share. Shalleen, would you mind being the scribe? A scribe is a person who writes for the group.
>
> *Shalleen:* [No response.]
>
> *Omar:* I wonder whether the box was taped?
>
> *David:* I wonder if the tape is really better?
>
> *Zinya:* I wonder if the box was smaller if it would have fallen and broken like it did?
>
> *Teacher:* [Notices Shalleen did not respond.] Omar and David, those are good questions. Shalleen, do you wonder why they used a gorilla?
>
> *Shalleen:* Yes.
>
> *Teacher:* OK. So, your wonder question can be, "I wonder if the gorilla is real."
>
> *Shalleen:* [Nods in agreement.]

Notice how the teacher understood Shalleen's level of speaking was at the beginner level and did not force her to answer. The teacher asked her a question, waited for Shalleen to respond, then phrased it for her so she could hear how to say it. This is intentional and an effective strategy when working with multilingual learners. The teacher also knew that Shalleen's writing score on the WIDA assessment was between emerging and developing, so Shalleen would be able to scribe for the group.

Pause for a moment and reflect on the following questions.

- Teachers need to develop a range of ways to interact with all the students in the class and engage with them. Where did the teacher demonstrate this interaction and engagement?

- What do you notice about the teacher's interaction with Shalleen? How did the teacher correct Shalleen's grammar without singling her out?
- How did the teacher involve Shalleen in the group activity based on her language proficiency level to keep her an active participant in the group?

Figure 6.9 connects the WIDA Can Do Descriptors with the Discuss purpose.

Discuss: Interacting with others to build meaning and share knowledge			
Beginner	**Intermediate**	**Advanced**	**Native or Fluent**
Students can: Use nonverbal behaviors to show engagement and listening, can answer "Wh-" questions, and can contribute by sharing own work	Students can: Support ideas with examples, ask clarifying questions, generate new questions, demonstrate awareness of personal bias, and defend point of view.	Students can: Build on others' ideas and listen with purpose (challenge others' ideas).	Students can: Present organized ideas and information on content, including use of graphics and multimedia; synthesize ideas of several speakers; and respond with evidence and examples.

Source: © 2016 by WIDA. Used with permission.

FIGURE 6.9: Discuss connection to WIDA Can Do Descriptors.

The teacher has the groups share their notice and wonder statements with the class. Since the teacher understands that beginner multilingual learners should not be forced to speak in front of the class, the teacher does not call on them unless they volunteer. It is important for the quads to share among themselves so that they become comfortable sharing their thoughts with the class as their language proficiency increases from beginner to advanced. The teacher then informs students that they will be looking at the commercial again.

> ***Teacher:*** *We will watch the commercial again for the third and final time, but this time I want you to think about what the advertisers want you to believe about the product. I have placed sentence starters on each group's table. Use these to help you make your statement.*

The students watch the commercial for the third time and discuss in their groups what they think the advertisers want them to believe. Meanwhile, the teacher gives each group the CER graphic organizer with sentence starters on it (figure 6.10).

> ***Teacher:*** *I have placed a graphic organizer on your table. Do not worry about that now. Are there any groups that want to share what they think is the commercial's purpose? What do you think the advertisers want you to believe?*
>
> ***Quad 4:*** *We think they want us to believe that if she used the Gorilla tape, the box would not have broken open.*
>
> ***Teacher:*** *[Asks a "Wh-" question to include the multilingual learner in the conversation.] Who thinks the advertisers want you to believe the tape is as strong as a gorilla? [Flexes muscles to represent strong.]*

Teacher: *Why do you think they used a gorilla and not a dog like a poodle? [The teacher shows a picture of a poodle.]*

---

1. I believe that the advertisers want us to _____ .

2. I believe _____ .

3. It is my opinion that advertisers used a _____ because they are tough and the _____ will be tough too.

4. I don't think using a _____ in the commercial means that the _____ is strong too.

5. Yes, I think so because _____ .

6. No, I do not think so because _____ .

FIGURE 6.10: Sentence starters.

Quad 2: *We think they want us to believe that the tape is as strong as a gorilla. Poodles aren't strong like a gorilla.*

Teacher: *[Once again to include the beginner multilingual learner, the teacher asks a "Wh-" question.] Who thinks that a poodle is weaker than a gorilla? [The teacher acts weak. For example, the teacher may act like it is a struggle to lift a lightweight object.]*

Teacher: *So, at the end of the commercial they state that "Gorilla packaging tape is the best for the toughest jobs on planet Earth?" Do you agree? Discuss with your group. I have placed sentence starters on your tables and on the board for you to use.*

[The teacher circulates around the room again to listen to the groups discuss the question and to ensure the multilingual learners are being included in the conversation. The teacher once again listens to Shalleen, Omar, Zinya and David's group.]

Teacher: *What do you think? Is Gorilla packaging tape the best for the toughest jobs on planet Earth?*

Omar: *I believe that the advertisers want us to believe the tape is the toughest.*

Shalleen: *Me too.*

David: *I think so, too, Omar. That is why they used a gorilla, which is a strong animal.*

Zinya: *I agree with everyone that the commercial is trying to make us believe that because they are using a gorilla, we are to think that means the tape is strong too.*

Teacher: *What in the commercial makes you think that?*

Shalleen: *[Shows a picture of a gorilla and the word* strong.*]*

Omar: *I agree. They show a box breaking and a strong gorilla, and the name of the tape has gorilla in it too.*

David: *I agree with both of you. They use a gorilla to make you think the tape is as strong as a gorilla is strong.*

Zinya: *I agree with everyone. The commercial is trying to trick us.*

Teacher: *There were some great conversations, and I'm glad to see many of you used the sentence starters. What we just did was the first step in the argumentation framework. The first step is to make a claim. You make a claim by answering a question. I asked a question, "Is Gorilla packaging tape the best for the toughest jobs on planet Earth?" Your response was the claim.*

Following this example, the teacher should review the graphic organizer and go over each section of the CER framework, modelling how it should be completed. It is important to state to the students that they will do it together many times, they should not expect to do it perfectly the first time, and it will take repetition before students are very good at it.

Model the CER Process

What might it look like to model the CER process for your class? In this section, let's break it down in a way that would be applicable for multilingual learners and their native English-speaking peers.

Teacher: *I want you to write the question in the section that says, "Question."*

Point out the question row to students, as shown in figure 6.11.

Example 1: TV commercial—Gorilla packaging tape

Question: *Is Gorilla packaging tape the best for the toughest jobs on planet Earth?*™

Gorilla Packaging Tape Commercial
www.youtube.com/watch?v=Hz12277Aynl

FIGURE 6.11: Question row.

Teacher: *Look at the next section where, in the first column, it has the word* claim. *It also has sentence starters for those of you who want to use them. We have used sentence starters many times. You use them the same way when writing a CER. Your claim will answer the question I just asked. Your claim must be written in a complete sentence. You cannot just write yes or no. Discuss with your group what* *your claim is. I will walk around to answer any questions that you may have if you are still unsure as to what to do.*

Point out the claim row to students, as shown in figure 6.12.

Recounting: To retell to display knowledge or narrate experiences or events (claim)			
Beginner	**Intermediate**	**Advanced**	**Native or Fluent**
Students reproduce words or phrases related to topics, work with partner to reach conclusions, and label illustrations.	Students are provided wait time to produce short paragraphs with the main idea and some detail. Sentence starters can be used if needed.	Students can reproduce a sequence of events or experiences using transitional words. They can summarize based on steps provided.	Students can provide concluding statements that support information provided, convey sequences, and show relationships of events.

FIGURE 6.12: Claim row.

Teacher: Now that everyone in the group has completed their claim, let's move to the next part of CER, the evidence.

[The teacher models their thoughts out loud so students can hear how to approach the process.]

Teacher: Let's see. My claim is that I do not believe that Gorilla packaging tape is the best for the toughest jobs on planet Earth. What evidence is there in the commercial that proves this? It can be words that the actors state, or something that I have seen in the commercial. I can also use the sentence starters to help me write my evidence statement or statements.

[The teacher displays the template on the projector for all students to watch while stating evidence aloud for the class.]

Teacher: Here is my evidence. The first evidence, based on the commercial, is that packing a box with notebooks and papers is not what I would consider one of the toughest jobs on planet Earth. The second evidence is that the commercial does not show that the tape can work on a box filled with rocks or heavier materials. The third evidence is that there is no proof that the thickness of the tape and the length of the tape makes the tape better.

With these data, I can now write my evidence statements. You should have a minimum of two evidence statements but three is better. These are my three evidence statements.

1. The evidence from the commercial lacks proof that it can handle the toughest jobs on planet Earth because it only uses a large box filled with light items.
2. The commercial does not show how well the tape would work on a box full of rocks.
3. The commercial lacks any evidence that the thickness and length make a difference in what the tape can do.

Point out the evidence row, as shown in figure 6.13 (page 174).

Explain: To clarify the why or how of ideas, actions, or phenomena (evidence)			
Beginner	**Intermediate**	**Advanced**	**Native or Fluent**
Students are provided sentence starters and word banks. They can compare illustrated descriptions and produce statements related to the main idea.	Students are provided wait time to produce short paragraphs with the main idea and some detail. They can connect content-related topics to the main idea.	Students can compare content-related ideas from multiple sources and produce information and text around graphs and charts.	Students can determine two or more central ideas in text. They can evaluate how ideas influence individuals or events and vice versa.

FIGURE 6.13: Evidence row.

Teacher: You have watched how I modeled the process. Now it's your turn to work in your groups and complete the evidence portion of the CER template using your group's claim. I will circulate the room to answer any questions and provide feedback.

[After students have completed their evidence portion of the CER template, the teacher moves on to the next part—creating the reasoning statement.]

Teacher: Now that I have cited my evidence, I need to use it to write a reasoning that supports my claim that the Gorilla packaging tape is not the toughest tape on planet Earth. The reasoning statement needs to be tied to my data, which, in this case, was what occurred in the video. Here is my reasoning statement.

Tough is an object that is strong enough to withstand adverse conditions or rough or careless handling. A tough job is where people push the limits all the time and it is a challenge to accomplish a goal. Gorilla packaging tape was not used in adverse conditions; rough, careless handling; nor used in a situation where people were pushing the limits all the time. Since it must have these characteristics to be considered tough, my evidence suggests that Gorilla packaging tape is not the best for the toughest jobs on planet Earth.

[The teacher repeats the process of circulating the room to answer questions and provide feedback. The teacher provides sentence frames for the multilingual learners at varying levels of language proficiency.]

Point out the reasoning row, as shown in figure 6.14.

Argue: To persuade by making claims supported by evidence (reasoning)			
Beginner	**Intermediate**	**Advanced**	**Native or Fluent**
Students can generate words and phrases like, "I think . . ." They can make a list of topic choices with peers and can connect simple sentences to form content-related ideas.	Students can substantiate opinions with content-related examples and evidence. They can provide feedback to peers on language used for claims and evidence.	Students can justify ideas using multiple sources and present opinions in reports backed by research.	Students can introduce claims and opposing claims along with reasons and evidence. They can create closing and concluding statements that support these claims.

FIGURE 6.14: Reasoning row.

Using the WIDA Can Do Descriptors (WIDA, n.d.c) is essential for teaching multilingual learners because they offer clear, practical insights into what students can achieve at different levels of English proficiency. These descriptors help teachers design lessons that are both challenging and supportive, tailored to the students' current language skills, even for topics as challenging as claim, evidence, and reasoning. By emphasizing what multilingual learners can do, educators can build on their strengths, boost their confidence, and promote meaningful language development. This approach fosters an inclusive classroom environment where multilingual learners can actively engage and succeed academically.

You will now meet Wang Xiu Ying, your new student from Shanghai, China. In the following sections, you'll see how CER is used for an actual challenge in the classroom. This challenge engages students in practical mathematics applications and enhances their critical-thinking skills by requiring them to use claims, evidence, and reasoning to make data-driven decisions.

Meet Wang Xiu Ying
https://youtu.be/mk0KRDu Pdkw?feature=shared

Meet Wang Xiu Ying

Meet your new student, Wang Xiu Ying. Figure 6.15 contains his bio and academic profile. Scan the QR code to view a message from Wang Xiu Ying.

Wang Xiu Ying

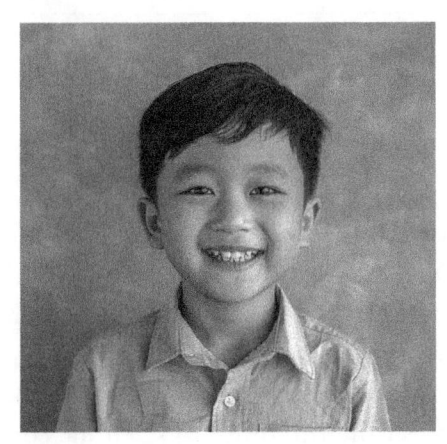

Wang Xiu Ying is eight years old from Shanghai, China. He lives with his mother, father, and grandmother. Shanghai is known for its rigorous education system, and it has prepared him well. The education system stands out as one of the strongest in the world. In his old school, Wang Xiu Ying was accustomed to long hours of study, starting his day with early-morning exercises and continuing with subjects like mathematics, Chinese, and science. His teachers emphasized discipline and excellence, often assigning homework to reinforce daily lessons.

As China's most westernized city, Shanghai has long been called *China's window* or *gate to the outside world* and the *bridge between East and West*. English is a required subject in China, and most students are taught their first lesson in the third grade, while others even start during kindergarten. Wang Xiu Ying was just beginning to learn English in his school before his father was relocated to the United States for work, but his father and mother spoke English a lot with him at home. They also worked with him on reading and writing so he would be prepared for English when he went to third grade.

Name	Country of Birth	Primary Language	Level of Language Proficiency	Age	Grade
Wang Xiu Ying	China	Chinese	Level 3: Developing	Eight	Third
WIDA Results					
Listening: 23.3		Speaking: 3.4	Reading: 2.8		Writing: 1.9
Student Capabilities By Level					

Listening (proficiency level 3):

Students can understand the main ideas and some details of spoken English in familiar contexts. They can follow multistep directions, especially when they are supported by visuals or gestures. They can comprehend simple stories, classroom instructions, and basic conversations. Students may need repetition or rephrasing to fully understand more complex or unfamiliar topics, but they are able to grasp the overall meaning and respond appropriately.

Speaking (proficiency level 2):

Students can participate in conversations on familiar topics using simple and compound sentences. They can ask and answer questions, share ideas, and describe experiences with some detail. While students can communicate effectively, their speech may include errors in grammar and pronunciation, and they might occasionally struggle to find the right words. Students are able to make themselves understood in most social and classroom interactions but benefit from support, such as visual aids or modeling, from peers and teachers.

Reading (proficiency level 2):

Students can understand general and some specific information in texts on familiar topics. They can identify the main idea and some details in short passages or stories with visual support. Students can follow simple written directions and use context clues to understand the meaning of unfamiliar words. They benefit from texts with pictures, diagrams, and other visual aids to support comprehension.

Writing (proficiency level 2):

Students can produce simple sentences and phrases related to familiar topics and personal experiences. Writing includes basic vocabulary and some common phrases, but it may lack grammatical accuracy and complexity. They can complete guided writing tasks with support, such as sentence starters or word banks. Their written work is often short and may include errors, but it conveys basic meaning.

FIGURE 6.15: Wang Xiu Ying's bio and academic profile.

CER STEM Challenge

This CER Challenge is a culmination of using both CER and the 5E model (chapter 5, page 138). Combining the 5E Model with the claim, evidence, reasoning (CER) framework enhances student learning by providing a structured, inquiry-based approach that promotes critical thinking, scientific literacy, and effective communication. This combination engages students in hands-on activities, guides them to formulate and support claims with evidence, and encourages them to connect their findings to scientific concepts. As students progress through the 5E phases, they deepen their understanding and apply their knowledge in new contexts, leading to a richer and more integrated learning experience. This method mirrors the practices of real scientists, fostering a more authentic and engaging educational environment. Figure 6.16 contains the CER STEM challenge lesson plan.

CER STEM Challenge Lesson Plan: The Great Pizza Debate		
Students will determine which pizza deal offers the best value for the money using claims, evidence, and reasoning.		
Engage	Show the students a picture of three pizzas as described in the three following choices. Tell the students that their parents agreed to let them have a pizza party. Their parents gave them the name of three pizzerias they can choose from, but they said that they have to decide which one to use and it has to be the cheapest one.	
Claim Question	Which of the three pizzerias do you think has the best pizza deal per square inch and why? 1. Pizzeria A: A large pizza (sixteen-inch diameter) for fifteen dollars. 2. Pizzeria B: Two medium pizzas (twelve-inch diameter each) for twenty dollars. 3. Pizzeria C: A family deal of three small pizzas (ten-inch diameter each) for eighteen dollars.	
CER	**Sentence Starter**	**Scaffolds for Wang Xiu Ying**
Claim: Student responses should be more than a yes-or-no answer, except for beginner multilingual learners. All other students should use complete sentences or short responses.	I believe that _____ pizza will be cheaper per square inch because _____ . It is my opinion, pizza _____ will be cheaper per square inch because _____ . Pizza _____ will be cheaper because _____ .	Since Wang Xiu Ying's proficiency level is a 3, he can communicate effectively, although his speech may include errors in grammar and pronunciation. Assist Wang Xiu Ying in the following ways. • Provide Wang Xiu Ying with a sentence frame. • Explain with visuals what a pizzeria is. • Explain square inch and inch, since China uses metric units. • Convert pizza diameter to centimeters. • Allow Wang Xiu Ying to use square centimeters, then show him how to convert back to square inches.
Explore	The students work to find the square inch cost of each pizza(s).	
Explain	The students show their work and provide an explanation for their answer based on their calculations.	
Evidence: Students may use words, numbers, graphs, symbols, data tables, or drawings when communicating their problem-solving strategy.	The pizza from pizzeria _____ costs _____ per square inch. According to my calculations, pizzeria _____ is (cheaper) or (more expensive) because _____ . Pizzeria _____ is a better deal because _____ . Pizzeria _____ is more expensive because _____ .	Since Wang Xiu Ying's writing is a level 2, he will need to use sentence frames or starters to help write his answers. Assist Wang Xiu Ying in the following ways. • Provide sentence frames without using *more than* or *less than*, since these grammatical phrases are often difficult for multilingual learners to comprehend.

Elaborate	Students state if the evidence supports their claim.	
Reasoning: Students justify their problem-solving method by explicitly communicating how the information given in the problem helped them decide on a strategy to employ, the mathematics skill(s) they learned that helped solve the problem, and the concepts they built on to solve the problem.	This proves my claim was correct because _____ . This supports my claim because _____ . This confirms that my claim was correct because _____ . My claim was incorrect because _____ . I did not think about _____ when making my claim.	Since Wang Xiu Ying proficiency level is a 3, he can communicate effectively although his speech may include grammar and pronunciation errors. Assist Wang Xiu Ying in the following ways. • Allow Wang Xiu Ying to communicate with his quad first and give him the option of sharing with the class if he wants to. • Allow Wang Xiu Ying to show his work but allow his shoulder or face partner to assist in the explanation.
Evaluate	Provide the answer to the problem after the students have compared their results. Have students compare their results with classmates and discuss any differences in findings. For quads that did not get the correct answer, have them discuss where they think they made the error. Have students reflect on the importance of using mathematical reasoning to make informed decisions in real-life situations. Evaluate students based on their accuracy in calculating areas and cost per square inch. Assess their ability to clearly articulate their claim, evidence, and reasoning in both written and oral presentations. Consider their participation in discussions and ability to justify their conclusions with logical arguments.	
Closure	Engage the class in a discussion about how CER helped them decide which pizza was the best deal. Allow students time to talk about how they came to a consensus as a quad. Have them rate themselves on how they worked together and how the CER model was beneficial for them or how it was not beneficial to them.	
Scaffolds for Wang Xiu Ying: Use rephrasing as needed since Wang Xiu Ying is at level 3 for listening. He is proficient in speaking but may need assistance with new vocabulary. Provide him with sentence frames to help him find other ways to articulate his thoughts. Allow Wang Xiu Ying to use visuals to interact if needed.		

FIGURE 6.16: CER STEM challenge lesson plan.

Key Takeaways

CER is especially beneficial for multilingual learners because it provides a clear and structured framework for developing their language and critical thinking skills, but it takes time and a lot of practice. This structured approach also encourages active participation and interaction in the classroom, helping multilingual learners build confidence and proficiency in using English in meaningful ways.

The critical thing to remember is that students will not master this skill overnight. Teachers will experience better student success if they allocate time daily to model the CER process when solving multistep word problems. It is strongly recommended that teachers allow students to use CER when working independently, in groups, and as a whole class. Ultimately, the end goal is for students to understand the framework so they can apply it in real-world situations. Mastering CER empowers multilingual learners to engage more deeply with the content, contribute to discussions, and achieve academic and social success. The following are some key takeaways from this chapter.

- It is important to let multilingual learners know what the format and expectations will be for the lesson and activities.
- Multilingual learners must receive constant opportunities where they can have discussions peer to peer and with the whole group. This is how they can gain access to academic language and engage in interdisciplinary practices.
- Sentence starters are crucial for multilingual learners so that their focus can be more on the CER process and content and not the writing.
- Sentence starters provide a clear structure to help third-grade students formulate and express their claims, support those claims with evidence, and explain their reasoning effectively.
- Vocabulary can be integrated into the lesson without necessarily being a vocabulary lesson.
- It is imperative, especially for multilingual learners' success, that the teacher models the CER process each time it will be used. Each time the teacher can use gradual release of responsibility, effective instructional strategies for multilingual learners, and effective questioning techniques to ensure and assess that the students are grasping the process of writing CERs.
- Teachers need to model the process to provide multilingual learners with a visual for how to think and process the information.
- Using CER in the science classroom can be particularly beneficial to multilingual learners because, as they progress, they will be able to make valuable contributions that are valued for the merit of their ideas regardless of social status or linguistic accuracy (Francis & Stephens, 2018).

Epilogue

District leaders, administrators, school-based leaders, and teacher education program developers must understand the importance of prioritizing present and future teachers' training and development in teaching multilingual students with limited skills in speaking, reading, writing, and listening in English. Supporting and developing a strong and diverse STEM teaching workforce can increase the number of students exposed to STEM careers, help reverse the cycle of poverty that many multilingual learners and underserved students presently reside in, and simultaneously create a more equitable science education for all students.

Many teachers, as noted throughout this book, have reported that they feel unprepared to teach the fastest-growing student population that is traditionally underserved in STEM-based courses. This lack of proper teacher training has normalized the belief that a lack of language proficiency correlates to lack of cognitive ability, which is not the case.

All teachers of STEM programs and content areas should adopt the following practices to support multilingual learners to successfully participate in STEM courses alongside their native English-speaking peers.

- **Culturally responsive teaching:** Teachers must receive extensive and ongoing professional learning about culturally responsive teaching. This must include recognizing implicit bias and addressing how it impacts our instruction and expectations of multilingual learners' ability to succeed in rigorous courses like STEM.

- **Leveraging multilingual learners' assets:** Seasoned and new teachers should recognize that multilingual learners bring value and experiences to the STEM classroom that can enrich and provide a global view to lessons. Leveraging the assets that multilingual learners bring to the classroom offers many benefits, such as enhanced cognitive skills, cultural diversity, and improved language development. Multilingual learners often display superior problem-solving abilities and creativity due to their multilingual experiences, which enriches the learning environment for everyone. Highlighting their cultural backgrounds promotes inclusion and mutual respect, preparing students for a globalized world.

- **Grouping:** Collaborative learning is a powerful strategy that gives multilingual learners opportunities to practice and improve their ability to read, write, speak, and listen to English in its proper context. Creating mini environments in the classroom will provide a safe space for multilingual learners to more easily take risks when learning English and the scientific language they need to be academically successful.

- **Scaffolding:** Both the teacher and the multilingual learner benefit when teachers apply scaffolds in instruction. The teacher's self-efficacy increases, and the learner's confidence improves as they begin to feel like a valuable contributor to classroom discussions and assignments.

- **Claim, evidence, and reasoning:** CER helps multilingual learners organize their thoughts and express their ideas clearly, using the academic language necessary for their academic progress. Through CER, multilingual learners practice forming logical arguments, backing them with evidence, and explaining their reasoning, which improves their understanding and communication skills. This structured approach also encourages active participation and interaction in the classroom, helping multilingual learners build confidence and proficiency in using English in meaningful ways. Teaching CER is not an easy task. It requires a lot of practice and patience, but the benefits far outweigh the effort needed to teach it.

I truly believe that all teachers want all students to succeed academically and socially, whether they are native English speakers or learning English for the first time. But without the proper training, preparation, and ongoing supports, an already monumental task can easily become an impossible one in the eyes of an untrained and undeveloped teacher.

The strategies provided in this book are not exhaustive but are intended to bring attention to the practices that must be well-supported by school and district leaders and must become a staple component of all teacher education programs regardless of the track a teacher wants to follow. Multilingual learners are a growing population, and all teachers should know how to meet their unique needs. It is our responsibility to ensure that the students who are participating in our schools' STEM programs are a true reflection of the diverse world in which we live.

References and Resources

Achieve, Inc. (2017, September). *Next generation science standards: High school by topic.* Washington, DC: Author. Accessed at https://nextgenscience.org/sites/default/files/HSTopic.pdf on June 3, 2024.

Akgün, M., & Akgün, İ. H. (2020). The effect of digital stories on academic achievement: A meta-analysis. *Journal of Education and Learning, 9*(6). https://doi.org/10.5539/jel.v9n6p71

Alimansyah, E. (2023, April 17). *The gamification cheatsheet: Key elements and best practices.* Accessed at https://bootcamp.uxdesign.cc/the-gamification-cheatsheet-key-elements-and-best-practices-d14cc3f08bad on May 24, 2024.

American Psychological Association. (n.d.). *Implicit bias.* Accessed at www.apa.org/topics/implicit-bias on February 19, 2024.

Aronson, B., & Laughter, J. (2016). The theory and practice of culturally relevant education: A synthesis of research across content areas. *Review of Educational Research, 86*(1), 163–206. https://doi.org/10.3102/0034654315582066

Attention Span. (n.d.). In *Merriam-Webster's online dictionary.* Accessed at www.merriam-webster.com/dictionary/attention%20span on June 11, 2024.

AVID Open Access. (n.d.a). *Accelerate learning by building on student assets.* Accessed at https://avidopenaccess.org/resource/accelerate-learning-by-building-on-student-assets on February 19, 2024.

AVID Open Access. (n.d.b). *Incorporate playlists into blended learning.* Accessed at https://avidopenaccess.org/resource/incorporate-playlists-into-blended-learning on February 29, 2024.

Ballantyne, K. G., Sanderman, A. R., & Levy, J. (2008). *Educating English learners: Building teacher capacity roundtable report.* Washington, DC: National Clearing House for English Language Acquisition. Accessed at https://files.eric.ed.gov/fulltext/ED521360.pdf on May 22, 2024.

Barone-Crowell, H. (2020). *Lack of preparation for mainstream teachers of English language learners.* [Master's thesis, The College at Brockport, State University of New York]. SUNY Open Access Repository https://soar.suny.edu/bitstream/handle/20.500.12648/4846/ehd_theses/1275/fulltext%20%281%29.pdf?sequence=1&isAllowed=y

Barrington, K. (2020, December 9). *The pros and cons of tracking in schools* [Blog post]. Accessed at www.publicschoolreview.com/blog/the-pros-and-cons-of-tracking-in-schools on February 19, 2024.

Bilash, O. (2011). *Krashen's 6 hypotheses.* Accessed at https://sites.ualberta.ca/~obilash/krashen.html on February 19, 2024.

Borgen Project. (n.d.). *Top 10 facts about girls' education in South Sudan.* Accessed at https://borgenproject.org/top-10-facts-about-girls-education-in-south-sudan on June 7, 2024.

Brain Balance. (n.d.). *Normal attention span expectations by age* [Blog post]. Accessed at www.brain balancecenters.com/blog/normal-attention-span-expectations-by-age on February 19, 2024.

Breiseth, L. (2015). *What you need to know about ELLs: FAQs.* Accessed at www.colorincolorado.org/article/what-you-need-know-about-ells-faqs on February 19, 2024.

Buxton, C. A., & Allexsaht-Snider, M. (2017). *Supporting K–12 English language learners in science: Putting research into teaching practice.* New York: Routledge.

Carver-Thomas, D., & Darling-Hammond, L. (2017, August). *Teacher turnover: Why it matters and what we can do about it.* Palo Alto, CA: Learning Policy Institute. Accessed at https://learningpolicyinstitute.org/media/174/download?inline&file=Teacher_Turnover_REPORT.pdf on June 24, 2024.

Castelán, J. (2023, October 16). *Elements of gamification: Key components to enrich the e-learning experience.* Accessed at www.iseazy.com/blog/elements-of-gamification on May 24, 2024.

Change the Equation. (2017). *Ending the double disadvantage: Ensuring STEM opportunities in our poorest schools.* Washington, DC: Author. Accessed at www.ecs.org/wp-content/uploads/CTE_STEM-Desert-Brief_FINAL.pdf on June 11, 2024.

Chowdhury, R. (n.d.). *Archimedes and the golden crown.* Accessed at www.longlongtimeago.com/once-upon-a-time/great-discoveries/archimedes-and-the-golden-crown on February 19, 2024.

Colorin Colorado. (2018, March 21). *How a structure and routine can help ELLs feel comfortable* [Video file]. Accessed at www.youtube.com/watch?v=vyFgjadOiz0 on February 19, 2024.

Continental Press. (2022, October 5). *What is the importance of picture books for ELL students?* [Blog post]. Accessed at www.continentalpress.com/blog/importance-picture-books-for-ell-students on June 7, 2024.

Cook, K., Pinder, D., Stewart, S., Uchegbu, A., & Wright, J. (2019, October 4). *The future of work in black America.* Accessed at www.mckinsey.com/featured-insights/future-of-work/the-future-of-work-in-black-america on February 19, 2024.

Cooper, A. (2021). *And justice for ELs: A leader's guide to creating and sustaining equitable schools.* Thousand Oaks, CA: Corwin Press.

Cornell, D. (2024, May 26). *15 schema examples (in learning psychology).* Accessed at https://helpfulprofessor.com/schema-examples on June 7, 2024.

Council for Exceptional Children. (n.d.). *English language learners.* Accessed at https://exceptionalchildren.org/topics/english-language-learners on December 11, 2022.

Cowan, N. (2014). Working memory underpins cognitive development, learning, and education. *Educational Psychology Review, 26*(2), 197–223. https://doi.org/10.1007/s10648-013-9246-y.

Cummins, J. (1984). *Bilingualism and special education: Issues in assessment and pedagogy.* San Francisco, CA: College-Hill Press.

Data USA. (n.d.). Aircraft pilots & flight engineers. [Data USA]. Accessed at https://datausa.io/profile/soc/aircraft-pilots-flight-engineers#ethnicity on July 3, 2024.

Davila, S. (2023, September). Teaching language to scientific standards. *Language Magazine,* Accessed at www.languagemagazine.com/teaching-language-to-scientific-standards-references on June 11, 2024.

de Haan, D. (2019). *Increasing the self-efficacy of general education teachers of ELLs* [Unpublished doctoral dissertation].

de Haan, D. (2022, May 3). *6 effective practices for teaching STEM to MLLs* [Blog post]. Accessed at www.tesol.org/blog/posts/6-effective-practices-for-teaching-stem-to-mlls on February 19, 2024.

Dehghanzadeh, H., Fardanesh, H., Hatami, J., Talaee, E., & Noroozi, O. (2019). Using gamification to support learning English as a second language: A systematic review. *Computer Assisted Language Learning. 34*(7), 934–957. https://doi.org/10.1080/09588221.2019.1648298

de Jong, E., & Commins, N. L. (n.d.). *How should ELLs be grouped for instruction?* Accessed at www.colorincolorado.org/article/how-should-ells-be-grouped-instruction on June 4, 2024.

Desautels, L. (2012, September 5). *Neuroscience of learning: Dr. Judy Willis hand-outs for School of Education Conference, "Thinking beyond the boundaries through social and emotional learning."* Accessed at https://revelationsineducation.com/neuroscience-of-learning-dr-judy-willis-hand-outs-for-school-of-education-conference-thinking-beyond-the-boundaries-through-social-and-emotional-learning on June 5, 2024.

de Souza, M., & Lee, J. (2017). Understanding Mexican immigrant students in American schools: A case study of two *preparatorias* in México. *Cogent Education, 4*(1). https://doi.org/10.1080/2331186X.2017.1387963

Dotson, M. J. (2001). *Cooperative learning structures can increase student achievement.* Accessed at www.kaganonline.com/free_articles/research_and_rationale/311/Cooperative-Learning-Structures-Can-Increase-Student on June 7, 2024.

Duran, L. B., & Duran, E. (2004). The 5E instructional model: A learning cycle approach for inquiry-based science teaching. *The Science Education Review, 3*(2). Accessed at https://files.eric.ed.gov/fulltext/EJ1058007.pdf on February 19, 2024.

Dweck, C. S. (2007). *Mindset: The new psychology of success.* New York: Ballantine Books.

Echevarria, J., Vogt, M., & Short, D. (2012). *Making content comprehensible for English learners: The SIOP model* (4th Ed.). Hoboken, NJ: Pearson.

eLearning Infographics. (2018, October 23). *Gaming the classroom: The art and science of game based learning* [Infographic]. Accessed at https://elearninginfographics.com/the-art-and-science-of-game-based-learning-infographic/ on June 11, 2024.

EL Education. (n.d.). *Collaborative culture: Routines.* Accessed at https://eleducation.org/resources/collaborative-culture-routines on February 19, 2024.

English Learners Success Forum. (n.d.). *Guidelines for improving science and engineering materials for multilingual learners.* Albuquerque, New Mexico: Author. Accessed at https://assets-global.website-files.com/5b43fc97fcf4773f14ee92f3/63583dfce1ea050576a1b335_ELSF_Science_Guidelines-02b.pdf on February 19, 2024.

Erdner, M. (2020, September 29). *Getting started with culturally responsive teaching.* Accessed at www.edutopia.org/article/getting-started-culturally-responsive-teaching on July 8, 2024.

Erduran, S., Ozdem, Y., & Park, J.-Y. (2015). Research trends on argumentation in science education: A journal content analysis from 1998–2014. *International Journal of STEM Education, 2*(1). https://doi.org/10.1186/s40594-015-0020-1

Evason, N. (2018). *Mexican culture.* Accessed at https://culturalatlas.sbs.com.au/mexican-culture/mexican-culture-communication on September 23, 2024.

Farah, K., & Barnett, R. (2021). *Teachers need more relevant PD options.* Accessed at www.edutopia.org/article/teachers-need-more-relevant-pd-options on May 22, 2024.

Ferlazzo, L. (2019, January 29). *Response: Cooperative learning can promote ELLs' academic oral language.* Accessed at www.edweek.org/teaching-learning/opinion-response-cooperative-learning-can-promote-ells-academic-oral-language/2019/01 on February 19, 2024.

Ferlazzo, L. (2022, May 17). *Crystal ball predictions: What will education for ELL students look like in 20 years?* Accessed at www.edweek.org/teaching-learning/opinion-crystal-ball-predictions-what-will-education-for-ell-students-look-like-in-10-years/2022/05 on May 20, 2024.

Fisher, D., Frey, N., & Almarode, J. (2023, May 5). *Scaffolding success.* Accessed at www.languagemagazine.com/2023/05/05/scaffolding-success on June 7, 2024.

Fleenor, S., & Beene, T. (2019). *Teaching science to English learners.* Irving, TX: Seidlitz Education.

Fletcher, G. (2016, April 11). *Modeling with mathematics through three-act tasks* [Blog post]. Accessed at www.nctm.org/Publications/TCM-blog/Blog/Modeling-with-Mathematics-through-Three-Act-Tasks on December 10, 2022.

Francis, D. J. (2006). *Practical guidelines for the education of English language learners: Research-based recommendations for instruction and academic interventions.* Houston, Texas: Center on Instruction. Accessed at www.centeroninstruction.org/files/ELL1-Interventions.pdf on June 7, 2024.

Francis, D., & Stephens, A. (2018). *English learners in STEM subjects: Transforming classrooms, schools, and lives.* Washington, DC: The National Academies Press. Accessed at https://nap.nationalacademies.org/read/25182/chapter/1 on February 19, 2024.

Fregni, J. (2021, April 13). *The fight to keep English learners from falling through the cracks.* Accessed at www.teachforamerica.org/one-day/top-issues/the-fight-to-keep-english-learners-from-falling-through-the-cracks on February 19, 2024.

Frey, W. H. (2018a). *Diversity explosion: How new racial demographics are remaking America.* Washington, DC: Brookings Institution Press.

Frey, W. H. (2018b). *The US will become 'minority white' in 2045, Census projects.* Accessed at www.brookings.edu/articles/the-us-will-become-minority-white-in-2045-census-projects on February 19, 2024.

Funk, C., & Parker, K. (2018, January 9). *Diversity in the STEM workforce varies widely across jobs.* Washington, DC: Pew Research Center. Accessed at www.pewresearch.org/social-trends/2018/01/09/diversity-in-the-stem-workforce-varies-widely-across-jobs/ on June 11, 2024.

Gibbons, P. (2007). Mediating academic language learning through classroom discourse. In J. Cummins & C. Davison (Eds.), *International Handbook of English Language Teaching* (pp. 701–718). New York: Springer.

Gibbons, P. (2015). Scaffolding language, scaffolding learning. In P. Gibbons (Ed.), *Scaffolding language, scaffolding learning* (2nd ed., pp. 13–14). Portsmouth, NH: Heinemann.

Gillespie, A. (2021, September 9). *What do the data say about the current state of K–12 STEM education in the US?* Accessed at https://new.nsf.gov/science-matters/what-do-data-say-about-current-state-k-12-stem on February 19, 2024.

Global Precipitation Measurement Mission. (n.d.a). *Geographical influences on climate teacher guide.* Accessed at https://gpm.nasa.gov/education/sites/default/files/lesson_plan_files/geographical%20influences/Geographical%20Influences%20-%20TG.pdf on June 7, 2024.

Global Precipitation Measurement Mission. (n.d.b). *Precipitation towers: Teacher's guide.* Accessed at https://gpm.nasa.gov/education/sites/default/files/lesson_plan_files/PrecipTowers%20-%20TG.pdf on June 7, 2024.

Goldhaber, D., Lavery, L., & Theobald, R. (2015). Uneven playing field? Assessing the teacher quality gap between advantaged and disadvantaged students. *Educational Researcher, 44*(5), 293–307. https://doi.org/10.3102/0013189X15592622

Hall, E. T. (1976). *Beyond culture.* New York: Knopf Doubleday.

Hallman, H. L., & Meineke, H. R. (2016). Addressing the teaching of English language learners in the United States: A case study of teacher educators' response. *Brock Educational Journal, 26*(1), 68–77. https://doi.org/10.26522/brocked.v26i1.478.

Halwani, N. (2017). Visual aids and multimedia in second language acquisition. *English Language Teaching, 10*(6): 53–59. https://doi.org/10.5539/elt.v10n6p53

Hamid Mahmood. (2016, September 8). *Wiliam on PPPB—Pose, pause, pounce, bounce.* [Video]. YouTube. [www.youtube.com/watch?v=TMBsTw37eaE].

Hammond, Z. (2013). *Ready for rigor: A framework for culturally responsive teaching.* Accessed at https://crtandthebrain.com/wp-content/uploads/READY-FOR-RIGOR_Final.pdf on June 25, 2024.

Hammond, Z. (2015). *Culturally responsive teaching and the brain: Promoting authentic engagement and rigor among culturally and linguistically diverse students.* Thousand Oaks, CA: Corwin Press.

Harper, A. (2019, February 20). *English language learners need equal access to STEM opportunities, report finds.* Accessed at www.k12dive.com/news/english-language-learners-need-equal-access-to-stem-opportunities-report-f/548679 on May 22, 2024.

Hernandez, A. (2022). Closing the achievement gap in the classroom through culturally relevant pedagogy. *Journal of Education and Learning, 11*(2). https://doi.org/10.5539/jel.v11n2p1.

Hill, J. (2023, July 3). *Storytelling in the ESL classroom: The importance of personal narratives for English language learners.* Accessed at https://readtheory.org/teachers-lounge/storytelling-esl-classroom on June 7, 2024.

Hofstede, G., Hofstede, G. J., & Minkov, M. (2010). *Cultures and organizations: Software of the mind* (3rd ed.). New York: McGraw-Hill Education.

Hollie, S. (2018). *Culturally and linguistically responsive teaching and learning: Classroom practices for student success.* Huntington Beach, CA: Shell Education.

IRIS Center. (2002a). *What do teachers need to know about students who are learning to speak English?* Accessed at https://iris.peabody.vanderbilt.edu/module/ell/cresource/q1/p02/#content on February 19, 2024.

IRIS Center. (2022b). *Providing instructional supports: Facilitating mastery of new skills.* Accessed at https://iris.peabody.vanderbilt.edu/module/sca/#content on February 19, 2024.

IRIS Center. (2022c). *Teaching English language learners: Effective instructional practices.* Accessed at https://iris.peabody.vanderbilt.edu/module/ell/#content on February 19, 2024.

Jackson, Y. (2015). *Foreword.* In Z. Hammond's (Ed.) *Culturally responsive teaching and the brain.* (pp. vi–vii) Thousand Oaks, CA: Corwin Press.

Jimenez, L. (2020, September 14). *Preparing American students for the workforce of the future. center for American progress.* Accessed at www.americanprogress.org/article/preparing-american-students-workforce-future on February 19, 2024.

Johansen, M. O., Eliassen, S., & Jeno, L. M. (2023). "Why is this relevant for me?": Increasing content relevance enhances student motivation and vitality. *Frontiers in Psychology, 14.* https://doi.org/10.3389/fpsyg.2023.1184804

Johnson, D. W., & Johnson, R. T. (1992). What to say to advocates for the gifted. *Educational Leadership, 50*(2), 44–47.

Johnson, D. W., & Johnson, R. T. (2019). Cooperative learning: The foundation for active learning. *IntechOpen.* Accessed at https://doi.org/10.5772/intechopen.81086.

Johnson, M. S., & Rodriguez, F. P. (2005). *The handbook for teachers who work with children of Mexican origin.* Accessed at https://people.uncw.edu/martinezm/Handbook/html/barriers.htm on February 19, 2024.

Johnston, M. (2022, September 29). *Diversity, equity, and inclusion (DEI) in aviation.* Accessed at https://calaero.edu/diversity-equity-inclusion-dei-aviation on February 19, 2024.

Jones, A. (2018, May 17). *Questioning: pose, pause, pounce, bounce* [Blog post]. Accessed at https://hertsandbuckstsablog.wordpress.com/2018/05/17/questioning-pose-pause-pounce-bounce on February 19, 2024.

Kagan, S. (n.d.). *Kagan structures to enhance student motivation.* Accessed at www.kaganonline.com/free_articles/dr_spencer_kagan/492/Kagan-Structures-Enhance-Student-Motivation on February 19, 2024.

Kagan, S. (2015a). *Kagan cooperative learning.* San Clemente, CA: Kagan Publishing.

Kagan, S. (2015b). *10 reasons to use heterogeneous teams.* Accessed at www.kaganonline.com/free_articles/dr_spencer_kagan/396/10-Reasons-to-Use-Heterogeneous-Teams on February 19, 2024.

Kagan, S., & High, J. (2002). *Kagan structures for English language learners.* Accessed at www.kaganonline.com/free_articles/dr_spencer_kagan/279/kagan-structuresforenglish-language-learners on June 7, 2024.

Katz, N. (2020, March 2). *State education funding: The poverty equation.* Accessed at www.future-ed.org/state-education-funding-concentration-matters on December 11, 2022.

Khong, T. D. H., & Saito, E. (2014). Challenges confronting teachers of English language learners. *Educational Review, 66*(2), 210–225. https://doi.org/10.1080/00131911.2013.769425

Klein, A. (2022, May 3). *Ditch those math worksheets. The case for teaching real-world problem solving in K–5.* Accessed at www.edweek.org/teaching-learning/ditch-those-math-worksheets-the-case-for-teaching-real-world-problem-solving-in-k-5/2022/05 on June 7, 2024.

KNILT. (2021, November 3). *Lesson 2: Content, process, and product.* Accessed at https://knilt.arcc.albany.edu/Lesson_2:_Content,_Process,_and_product on February 19, 2024.

Krashen, S. (n.d.). *Stephen Krashen and language acquisition.* Accessed at www.montgomeryschoolsmd.org/siteassets/district/curriculum/esol/cpd/module2/docs/krashenFINALtext.pdf on May 21, 2024.

Krashen, S. (2017). The case for comprehensible input. *Language Magazine.* Accessed at https://sdkrashen.com/content/articles/case_for_comprehensible_input.pdf on February 19, 2024.

Krashen, S. D., & Terrell, T. D. (1998). *The natural approach: Language acquisition in the classroom.* Hoboken, NJ: Prentice Hall.

Kris, D. F. (2018, May 15). *Why reading aloud to kids helps them thrive.* Accessed at www.pbs.org/parents/thrive/why-reading-aloud-to-kids-helps-them-thrive on June 7, 2024.

Language Magazine. (2018, October 22). *1 in 4 students is an English language learner: Are we leaving them behind?* Accessed at www.languagemagazine.com/2018/10/22/1-in-4-students-is-an-english-language-learner-are-we-leaving-them-behind on December 11, 2022.

LeadingLearner. (2019, September 8). *Avoiding teaching to the middle.* Accessed at https://leadinglearner.me/2019/09/08/avoid-teaching-to-the-middle-red19 on February 19, 2024.

Lee, J., & Seel, N. M. (2012). *Schema-based learning.* In N. M. Seel (Ed.) *Encyclopedia of the Sciences of Learning* (pp. 2946–2949). Boston: Springer.

Leins, C. (2019, August 21). *Cities struggle to prepare African Americans, Latinos for the future workforce.* Accessed at www.usnews.com/news/cities/articles/2019-08-21/cities-struggle-to-prepare-african-americans-latinos-for-the-future-workforce on February 19, 2024.

Lesaux, N. K. (2013, February 27). *Focus on higher-order literacy skills.* Accessed at www.educationnext.org/focus-on-higher-order-literacy-skills on May 20, 2024.

Lesley University. (n.d.). *Empowering students: The 5E model explained.* Accessed at https://lesley.edu/article/empowering-students-the-5e-model-explained on February 19, 2024.

Leung, W. M. V. (2023). STEM education in early years: Challenges and opportunities in changing teachers' pedagogical strategies. *Education Sciences, 13*(5), 490. https://doi.org/10.3390/educsci13050490

Levine, L. N., Lukens, L., & Smallwood, B. A. (2013). *The GO TO strategies: Scaffolding options for teachers of English language learners, k–12.* Accessed at www.cal.org/wp-content/uploads/2023/10/go-to-strategies.pdf on February 19, 2024.

Lin, M., & Bates, A. (2014). Who is in my classroom? Teachers preparing to work with culturally diverse students. *International Research in Early Childhood Education, 5*(1), 27. Accessed at https://files.eric.ed.gov/fulltext/EJ1151003.pdf on July 8, 2024.

Linquanti, R., Cook, H. G., Bailey, A. L., & MacDonald, R. (2016). *Moving toward a more common definition of English learner: Collected guidance for states and multi-state assessment consortia*. Washington, DC: Council of Chief State School Officials. Accessed at https://ccsso.org/resource-library/moving-toward-more-common-definition-english-learner-collected-guidance-states-and on December 11, 2022.

Lisa, A. (2019, February 11). *50 ways the workforce has changed in 50 years*. Accessed at https://stacker.com/business-economy/50-ways-workforce-has-changed-50-years on February 19, 2024.

Lombardi, J. D. (2016, June 14). *The deficit model is harming your students* [Blog post]. Accessed at www.edutopia.org/blog/deficit-model-is-harming-students-janice-lombardi on February 19, 2024.

Lopez, J. K. (2016). *Funds of knowledge*. Accessed at http://web.archive.org/web/20160426095921/http:/www.learnnc.org/lp/pages/939?style=print on February 19, 2024.

Lynch, M. (2016, April 21). *What is culturally responsive pedagogy?* Accessed at www.theedadvocate.org/what-is-culturally-responsive-pedagogy on February 19, 2024.

Lyon, G. H., Jafri, J., & St. Louis, K. (2012, Fall). *Beyond the pipeline: STEM pathways for youth development*. Accessed at https://niost.org/Afterschool-Matters-Fall-2012/beyond-the-pipeline-stem-pathways-for-youth-development on February 19, 2024.

Ma, Y. (2022). The effect of teachers' self-efficacy and creativity on English as a foreign language learners' academic achievement. *Frontiers in Psychology, 13*. https://doi.org/10.3389/fpsyg.2022.872147.

Magic School AI. (n.d.). Retrieved from https://www.magicschool.ai.

Main, P. (2022, March 10). *Independent learning: A teacher's guide*. Accessed at www.structural-learning.com/post/independent-learning-a-teachers-guide on February 19, 2024.

Maxwell, L. A. (2014a, June 17). *Most teacher preparation falls short on strategies for ELLs, NCTQ finds*. Accessed at www.edweek.org/teaching-learning/most-teacher-preparation-falls-short-on-strategies-for-ells-nctq-finds/2014/06 on February 19, 2024.

Maxwell, L. A. (2014b, August 19). *U. S. school enrollment hits majority-minority milestone*. Accessed at www.edweek.org/leadership/u-s-school-enrollment-hits-majority-minority-milestone/2014/08 on February 19, 2024.

McCarthy, J. (2018, January 10). *Extending the silence*. Accessed at www.edutopia.org/article/extending-silence on June 7, 2024.

McDonald, A. (2018, February 02). *Funds of knowledge*. Accessed at www.notimeforflashcards.com on June 7, 2024.

McGraner, K. L., & Saenz, L. (2009). *Preparing teachers of English language learners*. Washington, DC: National Comprehensive Center for Teacher Quality. Accessed at https://eric.ed.gov/?id=ED543816 on May 20, 2024.

McNeil, L. A., & Luft, J. A. (2021). *Using claim-evidence-reasoning (CER) in an undergraduate chemistry class: An exploration of CER construction and race gender and status* [Doctoral dissertation, University of Georgia.] https://esploro.libs.uga.edu/esploro/outputs/9949421027202959.

McVee, M., Silvestri, K., Shanahan, L., & English, K. (2017). Productive communication in an afterschool engineering club with girls who are English language learners. *Theory Into Practice, 56*(4), 246–254. https://doi.org/10.1080/00405841.2017.1350490

Medina, J. (2014). *Brain rules: 12 principles for surviving and thriving at work, home, and school*. Seattle, WA: Pear Press.

Middlecamp, C. (n.d.). *Students speak out on collaborative learning*. Accessed at https://archive.wceruw.org/cl1/cl/story/middlecc/TSCMD.htm on February 19, 2024.

Miller, E. C. (n.d.). *Next generation science standards: Offering equitable opportunities for ELLs to engage in science.* Accessed at www.colorincolorado.org/article/next-generation-science-standards-offering-equitable-opportunities-ells-engage-science on December 11, 2022.

Miller, E., Lauffer, H. B., & Messina, P. (2014). NGSS for English language learners: From theory to planning to practice. *Science and Children, 51*(5), 55–59.

Mitchell, C. (2017, March 7). *Schools are falling short for many English-learners.* Accessed at www.edweek.org/policy-politics/schools-are-falling-short-for-many-english-learners/2017/03 on May 20, 2024.

Mitchell, C. (2020, January 7). *U.S. schools see surge in number of Arabic- and Chinese-speaking English-learners.* Accessed at www.edweek.org/teaching-learning/u-s-schools-see-surge-in-number-of-arabic-and-chinese-speaking-english-learners/2020/01 on May 22, 2024.

MND Staff. (2019, December 3). *Mexican students place last in international math, reading and science tests.* Accessed at https://mexiconewsdaily.com/news/mexican-students-place-last on February 19, 2024.

Mohamed, N. (2024, March 14). *Every teacher is a language teacher: Strategies for supporting multilingual learners of English in the mainstream classroom.* Accessed at www.tesol.org/blog/posts/every-teacher-is-a-language-teacher-strategies-for-supporting-multilingual-learners-of-english-in-the-mainstream-classroom on May 21, 2024.

Moll, L. (2024, May 30). *Funds of knowledge video* [Video file]. Accessed at https://eclkc.ohs.acf.hhs.gov/video/collecting-using-video on June 7, 2024.

Moll, L. C., Amanti, C., Neff, D., & Gonzalez, N. (1992). Funds of knowledge for teaching: Using a qualitative approach to connect homes and classrooms. *Theory Into Practice, 31*(2).

Mora-Flores, E. (2011). *Connecting content and language for English language learners.* Huntington Beach, CA: Shell Education.

Morgan, C. (n.d.). *Important considerations for elementary classroom placement, seating, and grouping for English learner success.* Accessed at www.theallaccessclassroom.com/3-important-considerations-for-class-placement-seating-and-grouping-ell-students-in-the-classroom on February 19, 2024.

Moseley, C., Bilica, K., Wandless, A., & Gdovin, R. (2014). Exploring the relationship between teaching efficacy and cultural efficacy of novice science teachers in high-needs schools. *School Science and Mathematics, 114*(7), 315–325.

Muñiz, J. (2019, February 27). *Four reasons English learners lack access to quality STEM learning* [Blog post]. Accessed at www.newamerica.org/education-policy/edcentral/4-reasons-english-learners-lack-access-quality-stem-learning on February 19, 2024.

Najarro, I. (2023, October 27). *Is grouping English learners the right approach? What new research says.* Accessed at www.edweek.org/teaching-learning/is-grouping-english-learners-the-right-approach-what-new-research-says/2023/10 on February 19, 2024.

NASA. (n.d.a). Precipitation Education. Accessed at https://gpm.nasa.gov/education on July 8, 2024.

NASA. (n.d.b). *The water cycle.* Accessed at https://gpm.nasa.gov/education/water-cycle on February 19, 2024.

NASA Jet Propulsion Laboratory. (n.d.). *Educator guide: Precipitation towers: Modeling weather data.* Accessed at www.jpl.nasa.gov/edu/teach/activity/precipitation-towers-modeling-weather-data on July 8, 2024.

National Center for Education Statistics. (2022). *English learners in public schools.* Accessed at https://nces.ed.gov/programs/coe/indicator/cgf on December 10, 2022.

National Governors Association Center for Best Practices & Council of Chief State School Officers. (2010). *Common Core State Standards for mathematics.* Washington, DC: Authors. Accessed at https://corestandards.org/wp-content/uploads/2023/09/Math_Standards1.pdf on June 28, 2024.

National Research Council. (2000). *How people learn: Brain, mind, experience, and school* (Expanded ed.). Washington, DC: National Academies Press.

Nation's Report Card. (2019a). *NAEP report card: science. National student group scores and score gaps [Grade 4]*. Accessed at www.nationsreportcard.gov/science/nation/groups/?grade=4 on May 29, 2024.

Nation's Report Card. (2019b). *NAEP report card: science. National student group scores and score gaps [Grade 8]*. Accessed at www.nationsreportcard.gov/science/nation/groups/?grade=8 on May 29, 2024.

NGSS Lead States. (2013). *Next Generation Science Standards: For states, by state*s. Washington, DC: The National Academies Press.

Northern, S. (2019, August 30). *The 5 E's of inquiry-based learning*. Accessed at https://knowledgequest.aasl.org/the-5-es-of-inquiry-based-learning on February 19, 2024.

NYU Steinhardt. (2018, October 29). *An asset-based approach to education: What it is and why it matters*. Accessed at https://teachereducation.steinhardt.nyu.edu/an-asset-based-approach-to-education-what-it-is-and-why-it-matters on February 19, 2024.

Office for Civil Rights. (2018). *STEM course taking: Data highlights on science, technology, engineering, and mathematics course taking in our nation's public schools*. Washington, DC: Author. Accessed at https://civilrightsdata.ed.gov/assets/downloads/stem-course-taking.pdf on May 29, 2024.

OpenAI. (2024). ChatGPT (June 16 version) [Large language model]. https://chat.openai.com/chat.

Partnership for Reading. (n.d.). *Fluency: An introduction*. Accessed at www.readingrockets.org/topics/fluency/articles/fluency-introduction on February 19, 2024.

PBS LearningMedia. (n.d.). *Global Precipitation*. Accessed at https://oeta.pbslearningmedia.org/resource/buac17-68-sci-ess-globalprecipitation/global-precipitation on July 8, 2024.

PBS LearningMedia. (2022). *Funds of knowledge: Discover it*. Accessed at https://nhpbs.pbslearningmedia.org/resource/dsl22-sci-ets-fundsofknowledge-en/professional-development-funds-of-knowledge on February 19, 2024.

Prokopchuk, N. (2022). *Language learning in k–12 schools: Theories, methodologies, and best practices*. Accessed at https://openpress.usask.ca/languagelearningk12/front-matter/background/ on February 19, 2024.

Quinn, H., Lee, O., & Valdés, G. (2012). *Language demands and opportunities in relation to next generation science standards for English language learners: What teachers need to know*. Stanford, CA: Stanford University, Understanding Language. Accessed at https://ul.stanford.edu/sites/default/files/resource/2021-12/03-Quinn%20Lee%20Valdes%20Language%20and%20Opportunities%20in%20Science%20FINAL.pdf on June 7, 2024.

Quintero, D., & Hansen, M. (2017, June 2). *English learners and the growing need for qualified teachers* [Blog post]. Accessed at www.brookings.edu/blog/brown-center-chalkboard/2017/06/02/english-learners-and-the-growing-need-for-qualified-teachers on December 11, 2022.

Reeves, J. R. (2006). Secondary teacher attitudes toward including English-language learners in mainstream classrooms. *The Journal of Educational Research, 99*(3), 131–143. https://doi.org/10.3200/joer.99.3.131-143

Regional Educational Laboratory. (2019, July). *What is the research on the effectiveness or impact of culturally responsive teaching practices on student outcomes?* Accessed at https://ies.ed.gov/ncee/edlabs/regions/midatlantic/askarel_106.asp on February 19, 2024.

Reiss, J. (2005). *Teaching content to English language learners: Strategies for secondary school success*. White Plains: Pearson Education.

Robertson, K. & Ford, K. (n.d.). *Language acquisition: An overview*. Accessed at www.colorincolorado.org/article/language-acquisition-overview on March 21, 2024.

Rodriguez, A. J., & Bell, P. (2018, October). *Why it is crucial to make cultural diversity visible in STEM education.* Accessed at https://stemteachingtools.org/brief/55 on June 6, 2024.

Roe, K. (2019). Supporting student assets and demonstrating respect for funds of knowledge. *Journal of Invitational Theory and Practice, 25,* 5–13.

Roesler, M. K. (2022). *Supporting English language learners: Preparing content area teachers to promote academic achievement among culturally and linguistically diverse learner populations* [Doctoral dissertation, Grand Valley State University]. Scholarworks@GVSU https://scholarworks.gvsu.edu/gradprojects/122

Rosales, B. M. (2022, October 28). *Collaborative, hands-on trainings crucial for English language educators.* Accessed at https://edsource.org/2022/collaborative-hands-on-trainings-crucial-for-english-language-educators/680473 on February 19, 2024.

Rotermund, S. and Burke, A. (2021). *Elementary and secondary STEM education.* Accessed at https://ncses.nsf.gov/pubs/nsb20211 on June 24, 2024.

Ruble, L. A., Usher, E. L., & McGrew, J. H. (2011). Preliminary investigation of the sources of self-efficacy among teachers of students with autism. *Focus on Autism and Other Developmental Disabilities, 26*(2), 67–74. https://doi.org/10.1177/1088357610397345

Salter, A. F., & Renken, M. D. (2017). A review of the benefits of argumentation in the science classroom. *Georgia Journal of Science, 75*(1), 108.

Samson, J. F., & Collins, B. A. (2012, April). *Preparing all teachers to meet the needs of English language learners: Applying research to policy and practice for teacher effectiveness.* Washington, DC: Center for American Progress. Accessed at https://files.eric.ed.gov/fulltext/ED535608.pdf on February 19, 2024.

Santibañez, L., & Gándara, P. (2018). *Teachers of English language learners in secondary schools: Gaps in preparation and support.* Accessed at https://escholarship.org/uc/item/6c95c6bx on February 19, 2024.

Sass, T. R. (2015). *Understanding the STEM pipeline.* Washington, DC: National Center for Analysis of Longitudinal Data in Education Research. Accessed at https://files.eric.ed.gov/fulltext/ED560681.pdf on June 11, 2024.

Schiff, A. (2023, May 12). *Schiff, Booker, Larson, Takano, & Hayes reintroduce bicameral legislation to boost teacher compensation.* Accessed at https://schiff.house.gov/news/press-releases/schiff-booker-larson-takano-and-hayes-reintroduce-bicameral-legislation-to-boost-teacher-compensation on May 22, 2024.

Schmitt, N., Jiang, X., & Grabe, W. (2011). The percentage of words known in a text and reading comprehension. *The Modern Language Journal, 95*(1), 26–43. https://doi.org/10.1111/j.1540-4781.2011.01146.x.

Shah, J., Zarske, M. S., & Carlson, D. W. (2006). *Designing ways to get and clean water.* Accessed at www.teachengineering.org/activities/view/cub_earth_lesson3_activity1 on February 19, 2024.

Shenandoah County Public Schools. (n.d.). *WIDA levels and descriptors.* Accessed at www.shenandoah.k12.va.us/en-US/title-iii-43ede3fe/wida-levels-and-descriptors-d66f0c0d on June 25, 2024.

Shi, Q. (2017). English language learners' (ELLs) science, technology, engineering, math (STEM) course-taking, achievement and attainment in college, *Journal of College Access, 3*(2), Article 5.

Shinnawi, A. (2021, December 1). *Harnessing the power of storytelling to support migrant and immigrant students.* Accessed at www.edutopia.org/article/harnessing-power-storytelling-support-migrant-and-immigrant-students on June 7, 2024.

Short, D. J., Becker, H., Cloud, N., Hellman, A. B., & Levine, L. N. (2018, July 29). *The 6 principles for exemplary teaching of English learners.* Alexandria, VA: TESOL International Association.

Simon Fraser University. (n.d.). *Stages and symptoms of culture shock.* Accessed at www.sfu.ca/students/isap/explore/culture/stages-symptoms-culture-shock.html on February 19, 2024.

Skarin, R. (2022, September 28). *New national research raises teachers voices about instructional materials for English learners* [Blog post]. Accessed at www.elsuccessforum.org/blog/new-national-research-raises-teachers-voices-about-instructional-materials-for-english-learners on February 19, 2024.

Skepple, R. G. (2015). Preparing culturally responsive pre-service teachers for culturally diverse classrooms. *Kentucky Journal of Excellence in College Teaching and Learning, 12*(6), 57–59.

Slavin, R. E. (2014). Making cooperative learning powerful. *Educational Leadership, 72*(2).

Sousa, D. A. (2006). *How the brain learns* (3rd ed.). Thousand Oaks, CA: Corwin Press.

Sousa, D. A. (2010). *How the ELL brain learns.* Thousand Oaks, CA: Corwin Press.

Southern Oklahoma State University. (2021, December 21). *Why are visual tools important for helping ELL students learn English?* Accessed at https://online.se.edu/programs/education/med-curriculum-instruction/esl/help-ell-students-learn-english on February 19, 2024.

Spiegelman, M. (2020). *Race and ethnicity of public school teachers and their students.* Accessed at https://nces.ed.gov/pubsearch/pubsinfo.asp?pubid=2020103 on June 5, 2024.

Staake, J. (2021, June 30). *What is scaffolding in education? A guide for teachers.* Accessed at www.weareteachers.com/what-is-scaffolding-in-education on February 19, 2024.

Structural Learning. (2021, August 16). *Scaffolding in education: A teacher's guide.* Accessed at www.structural-learning.com/post/scaffolding-in-education-a-teachers-guide on July 8, 2024.

Suardi, M. (2011). *Output-based aid in Vietnam: Access to piped water services for rural households.* Washington, DC: The Global Partnership for Results-Based Approaches. Accessed at www.gprba.org/sites/default/files/publication/downloads/O-2012-001863_OBA_ENG_Web.pdf on February 19, 2024.

Sullivan, E. T. (2019, February 14). *Seven steps to ensure English learners aren't left out of STEM.* Accessed at www.edsurge.com/news/2019-02-14-seven-steps-to-ensure-english-learners-aren-t-left-out-of-stem on May 22, 2024.

Teed, S. (2020). *Making the water cycle accessible and relevant for English language learners.* [Master's thesis, University of Northern Iowa]. Accessed at UNI ScholarWorks https://scholarworks.uni.edu/grp/1447 on September 23, 2024.

Texas Education Agency. (2020, April 30). *TExES English as a second language (ESL) supplemental #154 preparation manual.* Accessed at https://tea.texas.gov/academics/special-student-populations/english-learner-support/final-esl-154-test-prep-manual-revised-4-30-20.pdf on June 7, 2024.

Tharp, R. (2000). *Teaching transformed: Achieving excellence, fairness, inclusion, and harmony.* New York: Routledge.

Top Hat. (n.d.). *Culturally responsive pedagogy.* Accessed at https://tophat.com/glossary/c/culturally-responsive-pedagogy on February 19, 2024.

Ullman-Shade, C. (2015). *Inquiry charts.* Jamaica Plain, MA: reDesign. Accessed at www.redesignu.org/wp-content/uploads/2020/10/Inquiry-Charts_reDesign.pdf on February 19, 2024.

Understood.org. (n.d.). *What is culturally responsive teaching?* Accessed at www.understood.org/en/articles/what-is-culturally-responsive-teaching on February 19, 2024.

U.S. Census Bureau. (2015, November 3). *Census bureau reports at least 350 languages spoken in U.S. homes.* Accessed at www.census.gov/newsroom/archives/2015-pr/cb15-185.html on February 19, 2024.

U.S. Climate Data. (2024). *Atlantic City weather averages and climate* [Atlantic City, New Jersey]. Accessed at www.usclimatedata.com/climate/atlantic-city/new-jersey/united-states/usnj0015 on July 3, 2024.

U.S. Department of State Bureau of Educational and Cultural Affairs. (n.d.). *STEM innovations and global competence.* Accessed at https://teacherexchanges.catalog.instructure.com/courses/stem-innovations-and-global-competence on February 19, 2024.

U.S. General Services Administration. (2023, October 26). *Official language of the United States.* Accessed at www.usa.gov/official-language-of-us on May 22, 2024.

Vasquez, V. (n.d.). *Lowering the affective filter for English language learners facilitates successful language acquisition* [Blog post]. Accessed at www.collaborativeclassroom.org/blog/lowering-affective-filter-facilitates-language-acq on February 19, 2024.

Vigeant, F. (2021, November 17). *Science for the next generation: Preparing for the new standards* [Blog post]. Accessed at www.knowatom.com/blog/science-for-the-next-generation on May 20, 2024.

Visible Learning. (2018, March 28). *Hattie ranking: 252 influences and effect sizes related to student achievement.* Accessed at https://visible-learning.org/hattie-ranking-influences-effect-sizes-learning-achievement on June 7, 2024.

Volman, M. & 't Gilde, J. (2021). The effects of using students' funds of knowledge on educational outcomes in the social and personal domain. *Learning, Culture and Social Interaction, 28.* https://doi.org/10.1016/j.lcsi.2020.100472

Vygotsky, L. S. (1978). *Mind in society: The development of higher psychological processes.* Cambridge, MA: Harvard University Press.

Wanasek, S. (2023, December 21). *4 classroom gamification elements and examples.* Accessed at www.classpoint.io/blog/4-classroom-gamification-examples on May 24, 2024.

Wang, S., Lang, N., Bunch, G. C., Basch, S., McHugh, S. R., Huitzilopochtli, S., et al. (2021). Dismantling persistent deficit narratives about the language and literacy of culturally and linguistically minoritized children and youth: Counter-possibilities. *Frontiers in Education, 6.* https://doi.org/10.3389/feduc.2021.641796.

Ward, B. A. (1987). *Instructional grouping.* Education Northwest. Accessed at https://educationnorthwest.org/sites/default/files/InstructionalGrouping.pdf on June 7, 2024.

Washington Office of Superintendent of Public Instruction. (2023, October). *Funds of knowledge toolkit.* Accessed at https://ospi.k12.wa.us/sites/default/files/2023-10/funds_of_knowledge_toolkit.pdf on June 7, 2024.

Watson, S, Miller, T. L., Driver, J., Rutledge, V., & McAllister, D. (2005). English language learner representation in teacher education textbooks: A null curriculum? *Education, 126*(1), 148.

WIDA. (n.d.a). *Can do descriptors.* Accessed at https://wida.wisc.edu/teach/can-do/descriptors on June 25, 2024.

WIDA. (n.d.b). *ELD standards framework.* Accessed at https://wida.wisc.edu/teach/standards/eld on May 24, 2024.

WIDA. (n.d.c). *Mission and history.* Accessed at https://wida.wisc.edu/about/mission-history on March 21, 2024.

WIDA. (2016). *Can do descriptors.* Accessed at https://wida.wisc.edu/teach/can-do/descriptors on October 10, 2024.

WIDA. (2020). *WIDA English language development standards framework, 2020 edition: Kindergarten–grade 12.* Madison, WI: Board of Regents of the University of Wisconsin System. Accessed at https://wida.wisc.edu/sites/default/files/resource/WIDA-ELD-Standards-Framework-2020.pdf on May 24, 2024.

Will, M. (2018, May 22). *Early-grades science: The first key STEM opportunity.* Accessed at www.edweek.org/teaching-learning/early-grades-science-the-first-key-stem-opportunity/2018/05 on February 19, 2024.

Will, M. & Najarro, I. (2022, April 18). *What is culturally responsive teaching?* Accessed at www.edweek.org/teaching-learning/culturally-responsive-teaching-culturally-responsive-pedagogy/2022/04 on February 19, 2024.

Williams, J. (2023, February 2). *The 5 stages of culture shock and how they impact travelers.* Accessed at https://thecureforcuriosity.com/the-5-stages-of-culture-shock on February 19, 2024.

Willis, J. (2011, April 14). *A neurologist makes the case for the video game model as a learning tool* [Blog post]. Accessed at www.edutopia.org/blog/neurologist-makes-case-video-game-model-learning-tool on June 6, 2024.

Willis, J. (2021, November 5). *How cooperative learning can benefit students this year.* Accessed at www.edutopia.org/article/how-cooperative-learning-can-benefit-students-year on February 19, 2024.

Wisconsin Department of Public Instruction. (n.d.). *Inquiry chart (I-chart).* Accessed at https://dpi.wi.gov/sites/default/files/imce/ela/bank/RL.IKI_Inquiry_Chart.pdf on February 19, 2024.

Wood, D., Bruner, J. S., & Ross, G. (2006). The role of tutoring in problem solving. *Journal of Child Psychology and Psychiatry, 17*(2), 89–100. http://dx.doi.org/10.1111/j.1469-7610.1976.tb00381.x

Yuan, H. (2017). Preparing teachers for diversity: A literature review and implications from community-based teacher education. *Higher Education Studies, 8*(1), 9. https://doi.org/10.5539/hes.v8n1p9

Zehr, M. A. (2011, May 3). *It's hard to get a handle on federal ELL policy.* Accessed at www.edweek.org/policy-politics/its-hard-to-get-a-handle-on-federal-ell-policy/2011/05 on February 19, 2024.

Zohar, A., & Dori, Y. J. (2003). Higher order thinking skills and low-achieving students: Are they mutually exclusive? *Journal of the Learning Sciences, 12*(2), pp. 145–181.

Zong, J. & Batalova, J. (2015, July 8). *The limited English proficient population in the United States in 2013.* Accessed at www.migrationpolicy.org/article/limited-english-proficient-population-united-states-2013 on February 19, 2024.

Index

NUMBERS

5E learning cycle, 138, 139, 140, 176

A

academic vocabulary, 117–118, 123. *See also* vocabulary
acceptance stage, 27, 28. *See also* culture shock
accountability, individual accountability, 81
achievement gap, barriers to closing, 12–16
adjustment stage, 27, 28. *See also* culture shock
admission requirements, 13–14, 16
advanced fluency stage, 30. *See also* language acquisition and proficiency
all about me questionnaire, 54–55
ancestral folk knowledge, 114
anticipation guides, 120
artificial intelligence (AI), 159
assessments. *See also* WIDA (World-Class Instructional Design and Assessment)
    funds of knowledge and instruction and, 114
    inquiry charts and, 141
    NGSS and, 3–4
    prior knowledge and, 119–121
asset-based approach, 113. *See also* leveraging student assets and building content knowledge through scaffolding; student assets
attention span, 35–37
avatars, 33

B

background knowledge, 103–104, 123. *See also* prior knowledge
barriers to closing the achievement gap, 12–16
beginning fluency stage, 30. *See also* language acquisition and proficiency
behavior, 45–47
bias, 47–48
blind spots, 47
brainstorming sessions, 120
buddy system, 83
Bybee, R., 138

C

California Aeronautical University, 69
Carlson, D., 104
Carver-Thomas, D., 15
chalk talk, 140, 141
challenges and gamification, 33
ChatGPT, 159
chunking, 81
claim, evidence, and reasoning (CER)
    about, 155
    in action, 163–175
    CER checklists, 162
    CER rubric, 161
    CER STEM challenge, 176–178

claim, evidence, and reasoning
    student template, 160–161
comparison between three-act
    mathematics tasks and, 158
four-part framework of, 156–158
Gorilla® packaging tape examples and,
    163–172
        implementing the CER process,
            168–172
        key takeaways, 178–179
        language proficiency and, 158–159
        meet Wang Xiu Ying, 175–176
        modeling the CER process, 172–175
        practice of, 182
        in your classroom, 159–163
classroom environments, 111
climatograms, 150–152
clustered groups, 82. *See also* collaborative
    learning groups; groups
cognitive load, 37, 67
cognitive processing, 37
collaborative learning groups.
    See also groups
        about, 77–78
        cooperative learning models and,
            78–84
        key takeaways, 109–110
        meet Jean Pierre, Linh, and Amihan,
            88–93
        in practice, 93–102
        STEM challenge, 102–109
        Two Pairs in a Quad and, 84–88
collaborative learning, practice of, 182
collectivism, 43, 44–45, 53, 75
comprehension hypothesis, 117
content knowledge. *See* leveraging student
    assets and building content knowledge
    through scaffolding
cooperative learning models
        about, 78–79
        benefits of, 81–83
        concerns about cooperative learning,
            79–81
        types of, 82–84
creativity and student assets, 116
critical thinking and STEM, 1–2

cultural archetypes, 44
cultural diversity, 28–29
cultural grouping, 84. *See also*
    collaborative learning groups; groups
cultural iceberg, 46–47
cultural relevance, 115
culturally responsive environments, 53–54
culturally responsive teaching
        about, 41–43
        culturally responsive approaches,
            practicing, 67–69
        culture and behavior, 45–47
        culture and relationships, 44–45
        five components of, 60
        implicit bias and, 47–48
        key takeaways, 74–75
        meet Maria, 48–51
        in practice, 52–58, 181
        STEM challenge, 58–74
culture and language framework, 62
culture shock, 27–28, 52
Cummins' quadrant, 98, 99

D

Darling-Hammond, L., 15
differentiated pathways, 115
differentiation and funds of knowledge,
    118
diversity
        cultural diversity, 28–29
        current challenges in STEM
            and, 2–3
        in the workforce, 19, 43, 68, 69

E

early production stage, 30. *See also*
    language acquisition and proficiency
effect size, 124
empowering multilingual learners
    through STEM education. *See also*
    multilingual learners
        about, 9–12
        barriers to closing the achievement
            gap, 12–16
        implications for schools and teachers,
            16–22
        key takeaways, 22–23

engineering design process, 65–66, 106–107
English learners. *See also* multilingual learners
    use of term, 2
    word problems for, 118
enrichment stations, 116
entry tickets, 121
equity and inclusion, 37

F

face partners, 85, 86. *See also* Two Pairs in a Quad
family and community funds of knowledge, 118
feedback
    gamification and, 33, 34, 35
    scaffolded instruction and, 38
flexible vocabulary journal, 127–128
flight, making the cultural connection to lessons about, 69–74
folk knowledge, 114
Francis, D., 13
Frayer model, 128
frustration stage, 27, 28, 52. *See also* culture shock
funds of knowledge and instruction
    about, 115–117
    academic vocabulary and, 117–118
    ancestral folk knowledge and, 114
    funds of knowledge, terms defined, 113
    prior knowledge and, 119–121
    tapping funds of knowledge, 118–119

G

gamification, 31, 33–35
genius hour, 116
Gibbons, P., 117
GLAD strategies, 141
global competence, 42, 43
goals
    gamification and, 34, 35
    sample goals worksheet, 65
    STEM challenges and, 63, 65–67
    teacher self-efficacy and, 18

gradual release of responsibility, 37
graphic organizers, 120
groups. *See also* collaborative learning groups; Two Pairs in a Quad
    creating grouping of students charts, 95–102
    group work, 87–88
    grouping, use of term, 77
    practice of, 182
    strategic partner support groups, 71–73

H

Hammond, Z., 44, 47
heterogeneous grouping, 83, 84. *See also* collaborative learning groups; groups
higher-order thinking, 123
homogenous grouping, 83. *See also* collaborative learning groups; groups
honeymoon stage, 27, 28. *See also* culture shock
hoopster activity, 71

I

implicit bias as a barrier, 47–48
individual accountability, 81
individualism, 43, 44–45, 53, 75
inquiry charts, 140–141
inquiry learning, 115
instructional strategies and scaffolds, 124. *See also* scaffolding
interdisciplinary learning, 115–116
interest groups, 83. *See also* collaborative learning groups; groups
intermediate fluency stage, 30. *See also* language acquisition and proficiency
introduction
    about this book, 6–7
    critical thinking and STEM, 1–2
    current challenges in STEM, 2–4
    my story, 4–6

J

Johnson, D., 80
Johnson, R., 80
journaling, flexible vocabulary journal, 127–128
just-in-time enrichments, 116

K

Kagan, S., 87
Kagan cooperative learning structures, 84
Khong, T., 12
Kris, D., 130
K-W-L charts, 121

L

language acquisition and proficiency. *See also* multilingual learners
    about, 29–31
    CER and, 158–159
    culture and language framework, 62
    scaffolded instruction and, 37
    stages of, 30
    teacher preparation and, 20
    teacher self-efficacy and, 17
    understanding language development and proficiency, 52–53
    WIDA and, 30–32
language production, 36
leaderboards, 33
learning objectives, 63, 65
leveraging student assets and building content knowledge through scaffolding. *See also* scaffolding; student assets
    about, 111–113
    funds of knowledge and instruction, 114–121
    key takeaways, 153
    meet Fatou, 121–122
    practice of, 181
    scaffolds in practice, 122–134
    STEM challenge, 144–153
    STEM challenge—introduction, 134–144
    terms defined, 113–114
long-term memory, 67

M

makerspaces, 116
making instruction applicable through culturally responsive teaching. *See* culturally responsive teaching
mathematics
    CER model and, 157, 158
    Common Core Mathematics Standards, 136–137
    distribution of students enrolled in high school mathematics classes, 11
    mindsets and, 33–34
    report on K–12 achievement in, 15
    teacher turnover and, 15
    three-act mathematics task structure, 156, 158
McGrew, J., 16
mentorship, 20
minilessons
    STEM challenges and, 137–138, 153
    STEM minilesson plan, 142–144
mixed groups, 82. *See also* collaborative learning groups; groups
mixed-age grouping, 84. *See also* collaborative learning groups; groups
motivation and gamification, 34
multilingual learners. *See also* empowering multilingual learners through STEM education
    about, 25
    CER and, 158–159
    current challenges in STEM and, 2–3, 4
    definition of, 26
    distribution of students enrolled in high school mathematics classes, 11
    factors unique to their experience, 27–38
    funds of knowledge and, 114
    growth of, 10–11
    identity formation and, 39
    key takeaways, 38–39
    multilingual learners' assets, 26–27
    science score comparison for, 10
    socioeconomic barriers and, 14–16
    STEM as a solution and, 38

N

NAEP report, science score comparisons, 10
Najarro, I., 45

narratives, 33
Next Generation Science Standards (NGSS), 3–4, 63

P

paper strip activity, 70–71
participation, 37, 81
peers
    peer observations, 20–21
    scaffolded instruction and, 37
    strategic partner support groups, 71–73
Phillips, J., 10
picture-word match, 124–126
PIES, 80–81
playlist options, 116
pose, pause, pounce, and bounce, 129–130
positive interdependence, 80–81
predictions and gamification, 34, 35
preproduction stage, 30. *See also* language acquisition and proficiency
prior knowledge. *See also* background knowledge
    activating and assessing prior knowledge, 119–121
    multilingual learners and, 113
    scaffolded instruction and, 37
process of learning, 87
processing time, 36
professional development, 17–18, 20, 21
professional networks and communities, 20
proficiency levels, grouping by, 82
project-based learning, 115

Q

QR codes for
    blind spots, 47
    culturally responsive teaching, 43
    meet Amihan, 93
    meet Fatou, 121
    meet Jean Pierre Léger, 88
    meet Linh Kha, 90
    meet Maria, 48
    meet Wang Xiu Ying, 175
    memory and the hippocampus, 78
    STEM resources, 6
    storytelling and the brain, 131
    student-to-student relationships, 83
    why airplanes fly, 71
    word problems for English learners, 118
    working memory, 67

R

racism, 67. *See also* culturally responsive teaching
Ready for Rigor framework, 61, 62
reentry shock, 27. *See also* culture shock
registers, 78, 123
relationships
    culturally responsive environments and, 53–54
    culture and, 44–45
    funds of knowledge and, 118
    implicit bias and, 47–48
rewards, 33
rigor
    cooperative learning and, 80
    mixed groups and, 82
rotating groups, 83. *See also* collaborative learning groups; groups
roto-copter activity, 71
Ruble, L., 16

S

Saito, E., 12
scaffolding. *See also* leveraging student assets and building content knowledge through scaffolding
    impact of, 182
    in practice, 122–134
    scaffolded ideas, 100
    scaffolded instruction, 37–38, 105, 106, 107
    selecting strategies to accelerate the learning process, 123–124
    sentence starters and, 108, 159
    strategies, incorporating scaffolding into, 124–134
Scaffolding Language, Scaffolding Learning, Second Edition (Gibbons), 117
science

NAEP report, science score comparisons, 10
Next Generation Science Standards (NGSS), 3–4, 63
sentence starters
    CER and, 168, 170, 179
    example of, 102, 132, 152, 169, 171
    as scaffolds, 108, 159
    using, 159, 164
Shah, J., 104
shoulder partners, 85, 86. *See also* Two Pairs in a Quad
silent period, 30. *See also* language acquisition and proficiency
simultaneous interactions, 81
skill building hypothesis, 117
skill development, 37
skill-specific groups, 84. *See also* collaborative learning groups; groups
slow reveal, 126–127
socioeconomic barriers, 14–16
speech emergence stage, 30. *See also* language acquisition and proficiency
spinning blimps activity, 71
standards
    Common Core Mathematics Standards, 136–137
    Next Generation Science Standards (NGSS), 3–4, 63
    planning the STEM challenge, 63, 65
STEM. *See also* empowering multilingual learners through STEM education
    impact on early grades, 21–22
    leaky STEM pipeline, 12, 14, 16
    leveraging student assets and, 111–112
    as a solution, 38
STEM challenges
    background knowledge and, 103–104
    CER STEM challenge, 176–178
    cultural connections and, 69–74
    culturally responsive approach and, 67–69, 73–74
    engineering design process and, 106–107
    evaluation of, 108–109
    goals and action steps for, 65–67
    introduction to, 104–106, 134–144
    lesson plan for, 145–148
    minilessons and, 138, 142–144
    objectives and standards for, 63, 65
    overview of, 135
    partner support groups and, 71–73
    planning the STEM challenge, 60–69
    sample STEM challenge, 64
    scaffolded instruction and, 105, 106
    scenarios for, 107–108
    STEM challenge and collaborative learning groups, 102–109
    STEM challenge and culturally responsive teaching, 58–74
    STEM challenge and leveraging assets and building content knowledge through scaffolding, 134–144, 144–153
    versus traditional science labs, 59
    visuals and, 70
    vocabulary and, 70–71
Stephens, A., 13
storytelling, 130–134
strategic partner support groups, 71–73. *See also* collaborative learning groups; groups
stress, impact of, 34
student assets. *See also* leveraging student assets and building content knowledge through scaffolding
    asset-based approach, terms defined, 113
    multilingual learners' assets, 26–27
    ways to build on, 115–116
student chart for listing geographical features and the effect on weather and climate, 149
student scenarios
    collaborative learning groups in practice, 93–102
    meet Amihan, 92–93
    meet Fatou, 121–122
    meet Jean Pierre, 88, 89–90
    meet Linh, 90–91

meet Maria, 48–51
meet Wang Xiu Ying, 175–176
student surveys and funds of knowledge, 118

T

targeted language, 123
task-based grouping, 83. *See also* collaborative learning groups; groups
teacher preparation
    achievement gap and, 12–13
    implications for schools and teachers and, 16, 18–21
    percentage distribution of teachers by race or ethnicity from 2017 to 2018, 19
teacher self-efficacy, 16–18
think-pair-share, 120
three-act mathematics task structure, 156, 158
three-dimensional learning (3-D learning), 4, 112
translators, 53
Two Pairs in a Quad. *See also* groups
    about, 72, 78
    collaborative learning groups and, 84–88
    concerns about cooperative learning, 79–80
    creating, 93–102
    explanations of quads 1 and 2, 94
    group work and, 87–88

U

understanding multilingual learners' unique needs. *See* multilingual learners
urban education, 3
Usher, E., 16
using claim, evidence, and reasoning to build language fluency. *See* claim, evidence, and reasoning (CER)
using collaborative learning groups to support language acquisition and sustain rigor. *See* collaborative learning groups

V

Venn diagram for writing similarities and differences between weather and climate, 150
visuals, 70, 118
vocabulary
    building academic vocabulary, 117–118
    flexible vocabulary journals, 127–128
    second language instruction and, 123
    STEM challenges and, 63, 70–71
voice and choice, 115

W

wait time, 36, 129
whole-class activities, 84
WIDA (World-Class Instructional Design and Assessment). *See also* language acquisition and proficiency
    about, 30–31
    Can Do Descriptors, 170, 175
    key takeaways, 39
    levels of language proficiency, 32
    Maria's WIDA score breakdown, 68
    understanding language development and proficiency, 52
Will, M., 45
word problems for English learners, 118
working memory, 67

Z

Zarske, M., 104
zone of proximal development (ZPD), 78, 84, 117, 124

Coaching for Multilingual Student Success
Karen Johannesen Brock with Margarita Espino Calderón
Understand how to intentionally involve instructional coaches in equipping teachers to successfully implement high-impact strategies that meet the unique learning needs of multilingual students. Gain tools and practices to design professional learning plans that incorporate targeted strategies with the support of an instructional coaching program.
BKG172

Coaching Teachers in Bilingual and Dual-Language Classrooms
Alexandra Guilamo
Gain the skills you need to coach teachers in bilingual and dual-language classrooms. In this practical guide, you will discover a proven process for creating a fair and effective observation and feedback cycle to help support teachers in this important work.
BKF918

Promises Fulfilled
Margarita Espino Calderón and Shawn Slakk with Hector Montenegro
Discover research-based strategies preK–12 administrators and teacher leaders can implement to effectively identify and support English learners. Each chapter ends with discussion questions readers should share with staff or team members to promote English learners' success schoolwide.
BKF774

What STEM Can Do for Your Classroom
Jason McKenna
Author and educator Jason McKenna offers examples, tried and tested classroom projects, and collaborative strategies in this innovative resource designed to open up STEM education for K–6 educators in exciting and expansive new ways.
BKG088

Cultivating Competence in English Learners
Margarita Espino Calderón and Lisa Tartaglia with Hector Montenegro
This research-backed guide offers evidence-based strategies core content teachers can use immediately to improve daily practice. The authors explore the importance of SEL application to the English learning process and how to connect essential instructional elements to cultivate active, engaged learners.
BKG001

Solution Tree | Press

Visit SolutionTree.com or call 800.733.6786 to order.

GLOBAL PD

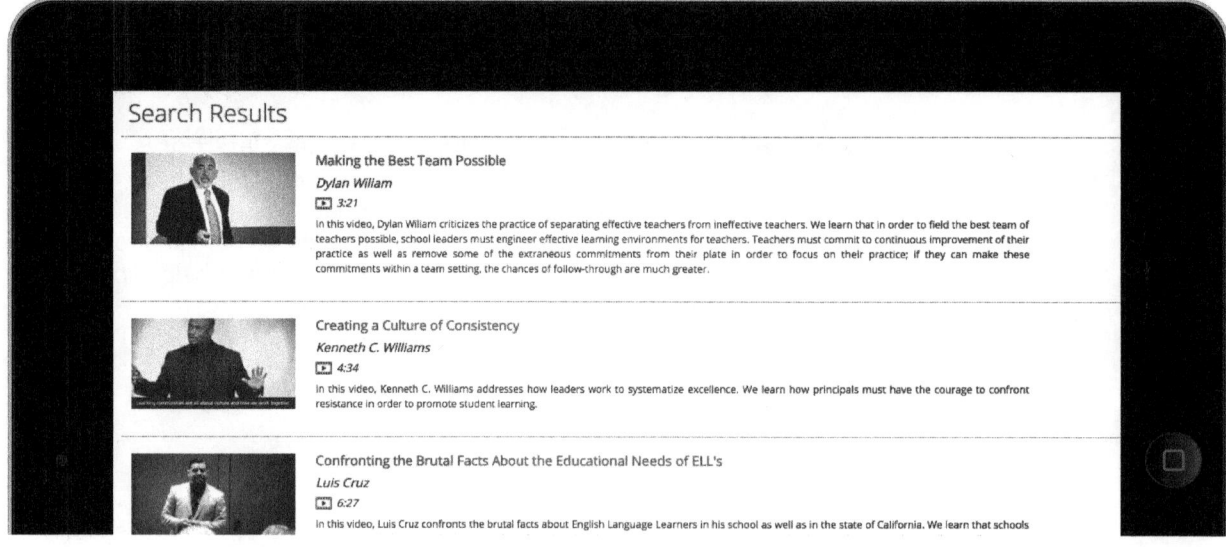

Access **Hundreds of Videos & Books** from Top Experts

Global PD gives educators focused and goals-oriented training from top experts. You can rely on this innovative online tool to improve instruction in every classroom.

- Gain job-embedded PD from the largest library of PLC videos and books in the world.

- Customize learning based on skill level and time commitments; videos are less than 20 minutes, and books can be browsed by chapter to accommodate busy schedules.

- Get unlimited, on-demand access—24 hours a day.

▶ **LEARN MORE**
SolutionTree.com/GlobalPDLibrary

 Solution Tree